Witness

Endangered Species of North America

For Jean Hiner Middleton, my mother and friend   S.M.

For my parents, and for Suzie Rashkis   D.L.

For Roy Eisenhardt   S.M./D.L.

Witness

Endangered Species of North America

by

SUSAN MIDDLETON and DAVID LIITTSCHWAGER

In Association with

The California Academy of Sciences

with text by
KEITH HOWELL
GORDY SLACK
BLAKE EDGAR

Chronicle Books San Francisco

Photography Copyright ©1994 by Middleton/Liittschwager
Text Copyright ©1994 by California Academy of Sciences/Middleton/Liittschwager

Excerpt from "Think Little" in *A Continuous Harmony: Essays Cultural
and Agricultural*, copyright ©1972 by Wendell Berry, reprinted by permission of
Harcourt Brace & Company.

Ezra Pound, *The Cantos of Ezra Pound* (excerpt).
Copyright ©1948 by Ezra Pound, reprinted by permission of
New Directions Publishing Corporation.

Printed in Japan

ISBN 0-8118-0258-2 (pb)   0-8118-0282-5 (hc)

Distributed in Canada
by Raincoast Books
112 East Third Avenue
Vancouver, B.C. V5T 1C8

10 9 8 7 6 5 4 3 2 1

Chronicle Books
275 Fifth Street
San Francisco, CA 94103

Book and Jacket Design by
Tenazas Design, San Francisco

Contents

Foreword

SUSAN MIDDLETON

What thou lovest well remains,
  the rest is dross
What thou lov'st well shall not be reft from thee
What thou lov'st well is thy true heritage

EZRA POUND, Canto LXXXI

THE PROCESSION OF 100 LIVES in the pages that follow is inspired by a desire to give endangered plants and animals a vivid presence in our human lives. As these species experience their homes, threatened ecosystems, disappearing around them, perhaps we, the human species, can understand ourselves as cause of their predicament and now their only hope. We, the photographers, offer these images as evidence of what is being sacrificed in the wake of our human transformation of the earth. It is our deep hope that if we practice love for these creatures, these kin of ours, they will remain with us as evidence of our true heritage.

We turned our eyes toward animals and plants, members of species with million-year histories, which may be the last of their kind. We looked, watched, and observed closely as we searched for the images we wanted. Each plant and animal is isolated. All other picture elements are removed to convey the beauty and uniqueness of each individual. We want the creatures to evoke a response in you who see them, that an intimate direct encounter might occur between you, the viewer, and the lives in these pictures as it did for us in our encounters with them in the photographic process. The absence of context, or any trace of habitat, bespeaks the precarious state of the species' homes. They are exposed and vulnerable as their habitats, the ecosystems to which they belong, are diminishing.

Besides attending to the individuals of a single species, we recognize the need to sustain their ecosystems. Yet we acknowledge the limits of photography to convey a whole ecosystem in a single photograph. So the emphasis of these photographs is on the wonder of an individual life. Profiles of the species are presented in the final section of the book. Here their habits and habitats, as well as the reasons for their decline and the steps taken to protect them, are described. We feel that an emotional connection to these animals and plants, most of which people will never have an opportunity to see, will increase the capacity for a wider perception of the diversity of life.

These species represent more than themselves and indicate a malaise that extends beyond them. They are the ones we can see; because of them we know there are many imperiled species that we cannot see. They stand for the over 800 North American plants and animals that are officially recognized as endangered, and which in turn point to the countless threatened species that have not made it through the federal bureau-

Hawaiian monk seal

Aleutian Canada goose

cracy. If endangered species are given refuge and their habitats are secured, they will act as umbrellas, protecting the many other animals and plants sharing their homes.

For us the project was a passport into the lives of these 100 endangered species. We approached them not as scientists but as photographers informed by science. A fascination with the diversity of life and concern for those life-forms threatened with extinction motivated our work. We learned about the life stories of our subjects, about the roles species play in their ecosystems, about biodiversity and evolution. We began to see forms and colors and textures not as ornamental, but as reactions to various influences, as brilliantly designed strategies for survival. This growing awareness deepened the revelation of beauty inherent in every plant and animal we photographed.

THIS NATIONAL PROJECT developed from a similar work of photographing endangered animals and plants in California between 1986 and 1990. In that work we realized that most of the species were unfamiliar to us and to most other people as well. Suspecting that what we discovered in California was true for all of North America, we proposed this project to photograph 100 North American endangered species listed as threatened or endangered by the United States government under the Endangered Species Act. Selected from the more than 800 species on the federal list, the group includes plants, insects, arachnids, crustaceans, mussels, fish, amphibians, reptiles, birds, and mammals. We chose species that live or once lived in the United States, but whose ranges collectively extend throughout North America. Each state is represented by at least two species whose ranges extend into that state. Areas where natural diversity is richest, such as the southeastern United States, Texas, California, and Hawaii, are represented with more species. In 1991, in affiliation with the California Academy of Sciences in San Francisco, we began the national project and immediately set to work to research, organize, and schedule fieldwork, projecting approximately two years to complete our work.

By presenting 100 endangered species together, we hoped to communicate the rich diversity of life in North America. We did not seek the most beautiful or dramatic species in the selection process. Most often we did not know what our subject looked like before we arrived to make the photograph. We wanted to include less celebrated species, which are as important to the functioning of the natural world as the more charismatic ones. The Madison Cave isopod, for example, does not have the allure of the bald eagle, but nevertheless it is a miracle of construction and adaptation. It is the product of millions of years of evolution in dark lakes deep underground inside Madison Saltpetre Cave, in West Virginia. Gradual adaptation of the species led to the loss of both its eyes and its body pigment, neither being of any use in an eternally nocturnal habitat. We learned that we might do well to be aware of this little creature, for its health is a measure of the quality of the groundwater on which people depend.

Some of the more obscure species proved to be the most captivating. When we arrived near Sumner Falls on the Connecticut River in New Hampshire to photograph the dwarf wedge mussel, we were apprehensive about photographing something that we presumed would look like a small black rock. A U.S. Fish and Wildlife Service biologist guided us to prime mussel habitat, and after hours of searching we prepared to photograph the only two mussels we were able to find. A small aquarium, specially designed for the mussel and built the night before, was set up next to our truck, which was parked at the river's edge. Water from the river was pumped through the aquarium while being aerated and filtered for clarity.

Searching for the Delta green ground beetle

Bald eagle blind

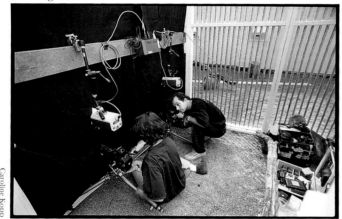

For the first half hour not much happened that we could see. Then, very gradually, the halves of the shell parted slightly and siphons began to emerge. The siphons are fringed, tubelike organs that draw in and eject water, filtering it for food. The mussel extended its foot into the sand we had placed in the aquarium, and with remarkable leverage it began to move about and dig itself into the sand, leaving only its siphons visible.

We photographed this process and then decided we wanted to see the whole creature. After consulting with the biologist, we removed the substrate and waited to see if the mussel would open once again. We were transfixed watching the mussel hold its balance and slowly open its siphons wider and wider, revealing some of its innermost parts. What we saw was strangely beautiful. After exposing several rolls of film, we returned the mussels to their bucket of river water and took them back to the spot where we had found them.

We were surprised to learn that each mussel establishes a long relationship with a particular stretch of riverbed, which can last up to 20 years. As filter feeders, freshwater mussels are acutely sensitive to water quality, and their decline indicates the degradation of the rivers where they live.

Although each is fascinating in its own right, many species are not immediately photogenic. In early August 1992, we drove 1,700 miles from St. Louis, Missouri, to Allagash, Maine, just across the Saint John River from Canada, to photograph a Furbish lousewort in bloom. A botanist for the Maine Natural Heritage Program met us and guided us to the plants. This was her first visit to the site that year and she doubted whether she could find them. There had been a serious ice scour on the Saint John River the previous spring, with ice jams carving the riverbanks into new configurations and carrying away some houses in the process. During a day of searching we found plants at two locations, flagging choice ones for further consideration.

The next morning we made several trips hauling our equipment through the forest and down the steep riverbank. Black flies, no-see-ums, and mosquitoes all took advantage of two succulent human beings. We surveyed the plants we had flagged and deliberated about how to make the picture. The Furbish lousewort is quite subtle in appearance, with small yellowish brown flowers. We had heard it described as rather homely and not much to look at. Because it is the only endemic plant in this postglaciated region—with no relatives anywhere nearby—it is an evolutionary mystery and has aroused the curiosity of botanists, many of whom have visited the site to study the plant firsthand. Most people living in the area have never seen the plant, don't know what it looks like, and are surprised at the attention it receives. Those who have seen it are even more puzzled by all the fuss. The majority of visitors to the region come to hunt bear, moose, or deer and to fish. There was a sign at Gardner Sporting Camp, where we stayed, that read "Guns, Ammo, and Home of the Furbish Lousewort."

On the riverbanks where it grows, the lousewort is surrounded by far showier plants, but none of them are as rare and unusual as the lousewort. Upon closer observation we saw that the brownish flowers are actually a mixture of yellow and deep burgundy red and that the stem is partially red where it is exposed to the sun. We became fond of the lousewort because it demanded that we stretch our vision to discover its distinctive beauty and character.

There are some endangered species that, by their very nature, command awe and respect. These "charismatic mega-vertebrates," as they are often called, such as the grizzly bear, Florida panther, red wolf, and wood bison, presented special challenges due to their large size and potential threat. With each of these species we worked with people who know the animals extremely well and took care to ensure our safety as well as that of the animals.

We will always remember photographing Big Guy, a male Florida panther, at White Oak Conservation Center in Jacksonville, Florida. He had been hit by a car when young and taken into captivity. Although now familiar with people, he was still essentially wild. After we patiently waited two days for this majestic animal to saunter in front of our black velvet background, he suddenly sat down in the ideal spot. We were inside his pen ready for this moment. Intensely and quietly, we began to photograph, advancing little by little to get the close-up we wanted. When just over 10 feet away, we asked the panther's keeper, "May we move the light three or four inches closer?" and he replied calmly, "Three, not four." We froze when suddenly the panther raised his head in a defensive posture, warning us not to come any closer. We were at the edge of Big Guy's territory and the keeper knew it. The panther was in charge, and we were all taking our cues from him. Tension filled the air as the panther looked straight into the lens for the few minutes we needed to expose the film. Then it was time to leave, quickly and softly.

Of all the animals we photographed, the black-footed ferret was the most protected and the least accessible. Under strict quarantine, with human contact kept to an absolute minimum, over 300 black-footed ferrets live together in a successful captive-breeding program in Wheatland, Wyoming, managed by the state's Game and Fish Department. Black-footed ferrets were once thought to be extinct, but were rediscovered in 1981. Careful monitoring of the wild population led to the conclusion, in 1986, that the only hope for the species was to capture the 18 surviving individuals for captive breeding and release back into the wild. The ferrets' situation was similar to that of the California condor, where the threat of extinction was deemed to be greater in the wild than in a captive breeding program.

The ferrets are highly susceptible to canine distemper and possibly human influenza, so they are kept in an immaculately clean facility. Each enclosure is either four by eight feet or four by four feet, raised three feet off the floor, and fitted with plastic tubing that forms tunnels leading down to breeding boxes. We sterilized all of our equipment before entering, took showers, and donned freshly cleaned jumpsuits, surgical masks, and rubber sandals. We set up our studio within one of the ferret enclosures, fastening the black velvet background inside and arranging lights outside the wire screening. We worked quietly and quickly, trying to minimize our presence. Surprised by the ferrets' loud screeching barks, we gasped with delight when the animals first poked their heads out of the tunnels in nearby enclosures. They are simultaneously endearing and fierce. In the wild, the menacing bark would surely startle a predator long enough to give a ferret a chance to flee.

Once we were ready to photograph, our subject was allowed to enter the enclosure. We were introduced to Cut-Lip, one of only two wild-caught males, a handsome and genetically valuable ferret. There were no guarantees that Cut-Lip would cooperate, and ultimately any success would depend entirely upon him. Puzzled at first by the new interior of his enclosure, Cut-Lip then gave us about 15 minutes of his attention before he walked

Presidio manzanita

Dwarf wedge mussel

over to the small hatch to be let down into his tunnel. Later on we tried again, and the second session went equally well. The next day we replaced the black background with white paper, and had two short sessions exposing black-and-white film. It was a privilege to observe these remarkable animals, knowing how perilously close they came to extinction, even though they were once widely distributed throughout North America.

One of the most fascinating and zaniest adventures was photographing a Virginia big-eared bat near Elkins, West Virginia. We had arranged to accompany a wildlife biologist and his assistant, both with the West Virginia Division of Natural Resources, in a field research and population study. Little did we know what awaited us.

We set out in two vehicles packed with gear for capturing and monitoring bats, as well as with photographic equipment. At a motel not far from the bat cave, we unloaded the photographic equipment and rearranged the furniture for a makeshift studio and lab. After dark we drove to the site. Packing bat cages and capture gear, and carrying flashlights, we traversed a stream on foot and hiked up a steep rocky mountain to the cave where the bats live. It was a hot night, but the breeze blowing from the cave was surprisingly chilly. Mosquitoes buzzed our ears as we helped position across the small cave entrance a three-foot-square aluminum frame rigged with a taut monofilament grid. The bats' echolocation fails to detect the thin strands of monofilament. When the bats attempt to exit the cave for their evening feast of insects, they are obstructed by the grid and fall into a holding pouch attached to the bottom of the frame. We stretched wads of nylon netting over gaps between the grid and the mouth of the cave to prevent any escapes. We waited several hours before we caught any big-ears, and by then it had started to rain. Distorting their echolocation, the rain drove the bats back to the cave for shelter. Bats were flying in all directions, often brushing next to our ears. Despite their density and proximity, as they dodged in and out of tight spaces, even between our legs, they never once touched us. Their speed and aerial mastery were impressive. Bats know better than to get caught in your hair.

Eventually we captured 11 big-ears, then dismantled our gear, loaded up, and descended the slippery mountain, bats in hand. Once back in the motel room, we photographed the bats until dawn, while the biologist and his assistant banded, weighed, and measured them, took hair samples, and recorded data. We fabricated a device we called a bat swivel, which was a rock attached to heavy wire armature drawn through the top of our clear plastic studio box. This provided the bat with something to grab onto and, when turned, gave us the opportunity to photograph from various angles. What spectacular creatures! With huge long ears that they periodically crumple up, they have sweet faces and attentive, shiny black eyes. The biggest challenge was to keep our subjects awake, with eyes wide open and ears perked up. We made various sounds to encourage them. Early in the morning the bats were taken back to their cave and released.

WE MOST OFTEN WORKED IN ZOOS, aquariums, living museums, captive-breeding and rescue facilities, botanical gardens, and theme parks, or with individuals who have permits to keep endangered species. In other words, we photographed species in exile from their natural homes. Many endangered species survive only as refugees, and those that are still able to live in the wild find their natural habitats diminishing. Often we were reminded of the difficult lives of endangered species. Nowhere was this more evident than when working with a captive-bred red wolf in a breeding facility in Graham, Washington. She was soon to be released in the Smokey Mountains in Tennessee as part of a successful reintroduction program. To heighten her chances for survival in the wild, she had been conditioned to be afraid of people. While caring for her, her keepers could not be friendly; they had to reinforce her natural fear of humans.

Many of our subjects were born wild but were injured and then rescued. With missing limbs or wings, pins in their joints, or other impairments, these animals could never survive in the wild. They live in zoos or in the loving hands of individuals permitted to care for them. Some of the species in this book that belong to this group are the Hawaiian hoary bat, wood stork, northern spotted owl, red-cockaded woodpecker, Florida panther, bald eagle, Hawaiian hawk (i'o), West Indian manatee, Aleutian Canada goose, and Florida Key deer. We photographed a Key deer in the Florida Keys that had been hit by a car. One of her legs was amputated and two others were broken. She was being cared for by United States Fish and Wildlife biologists who hoped to nurse her back to health. They called her Donna, short for prima donna, because she had charmed them into gathering her favorite foods, including bougainvillea.

Florida Key deer

We preferred photographing species in exile because it was less intrusive than working with species in their natural habitats. When we were unable to work with captive species, we joined population studies in which animals were captured temporarily, scientific data gathered, and the animals released. A few times we worked in the wild, setting up a back-ground and waiting for the animal to position itself in front of it. Through experience we learned that this never works without an incentive. The piping plovers were eager to return to their nest on the beach, and the Florida scrub jays were interested in peanuts when acorns were scarce during the winter. In both these instances we worked closely with biologists who were very protective of the animals. We often photographed plants in their natural habitats, usually remnants of once widespread, healthy wilderness. We photographed an eastern prairie fringed orchid in a narrow strip of native prairie sandwiched between a highway and a cornfield. Frequently we were in small preserves surrounded by human development. In Hawaii we photographed plants that grew inside fenced enclosures to protect them from feral animals.

Occasionally we had the good fortune to work in functioning ecosystems, elegant places where plants and animals have evolved together and established countless reciprocal relationships over immense periods of time with relatively little human disturbance. Here the flora and fauna are quintessential expressions of the place where they live. Organisms successfully adapting to their environment and to each other produce a dynamic

equilibrium, delicately balanced and constantly changing. Arising from this process of subtle, persistent struggles are plants and animals of unimaginable ingenuity and beauty, living in a community of finely tuned interactions. Intact ecosystems are rare—less than 5 percent of the earth's surface is legally protected from human encroachment on natural evolution. We experienced such places at the Florida scrub habitat at Archbold Biological Station, at Ash Meadows in southern Nevada, and in parts of Hawaii.

THE PROJECT HAS BEEN A PROCESS OF DISCOVERY. We hope the outcome will be a process of discovery for you. Our first look at an animal or plant was always an exciting moment. We were challenged, often amused, and dazzled by what we saw before our camera. We have learned that every living thing is astonishing in its own right. Many times we felt anxious as we searched for how best to convey the life we were observing. There were technical considerations, time constraints, physical limitations, and decisions about what aspects of a particular animal or plant we wanted to show in the photograph. Often hours, even days, would pass in this process; sometimes there was an immediate recognition of how to proceed.

When we saw the adult northern aplomado falcon and were then shown chicks in progressive stages of development, we knew immediately that we wanted to make a series of images that would be presented together. With the MacFarlane's four-o'clock, a striking plant belonging to the bougainvillea family, we didn't see the right picture until the end of a long day observing and photographing the plants on a steep hillside. Almost ready to pack up, we discovered a plant nestled in among others with a lovely lyrical curve to its stem. It had healthy leaves, mature flowers, and buds. Making a long day even longer, we worked on until the sun set.

The discovery process did not end with taking the photograph. In the darkroom, making the first prints, we saw new things. We could contemplate the intricate detail of feather patterns, the caviarlike configuration of tiny pollen sacs, the pupil and iris of an eye, and the expressions of plants as well as animals.

THIS GROUP OF PHOTOGRAPHS is in no way conclusive. They are not exact descriptions of these animals and plants, as in the tradition of scientific illustration. We approached our subjects not only as valuable specimens but also as living expressions of a spiritual presence in the world. Collectively, these photographs are meant to draw attention to our relationship as humans to the rest of the community of life on earth. The images are transmissions of some of the wonder we experienced when we made contact with each one of these plants and animals.

Every living organism has its own complicated way of perceiving and responding to what surrounds it. We, the human species, are part of this intricate web of life, woven over 3 billion years. At 2 million years old as a species, we are relative newcomers to the continuum of life on earth. We did not arrive as strangers. We came into being intimately connected with our natural environment: we depended on it for pleasure, meaning, and survival. In the course of our own evolution, these connections have become less direct and tangible. We have become estranged from the rich profusion of life around us, the life cycles of other creatures, and the fascinating dramas that take place among them. Increasingly this knowledge has been relegated to the realm of science, distant from the everyday life of people who are not biologists or naturalists.

We have managed to seriously upset nature's equilibrium. The extinction of species has accelerated to a rate unparalleled since the last great extinction—including the large dinosaurs—over 65 million years ago. At this rate we are imperiling life on earth as we know it. Science has revealed through the study of evolution and ecology the interconnectedness of all living organisms, and biologists are telling us that the hour is late for awakening to this catastrophic loss of life. It is becoming apparent that a solution to the preservation of biodiversity requires seriously limiting increases in human population. For only when we limit our own numbers and our consumption of natural resources will the long-term conservation of other species have a chance of succeeding.

Traversing this country to meet and photograph these fellow creatures became a kind of pilgrimage. In making this journey we have come to view humanity as one species among many and to embrace a less human-centered point of view. We humans are part of a life process bigger than ourselves. It is exhilarating to recognize ourselves as part of this magnificent process. The variety of life with which we share the earth will never be greater than it is at present. Sadly, it is in rapid decline. Life is grounded in biological diversity, and the fate of this diversity, which created and sustains us, is now in our hands. It will rest upon whether we love life enough to save it.

We belong to the natural world; it does not belong to us. The interconnectedness of life, in the present instant as well as through time, illuminates our place and our rightful relationship with other species here on earth. Can we find our place in the world as human beings if we devalue the rest of life? It is part of who we are. Can we hold out in our human citadel, separated from the remainder of life, and find any real meaning?

As photographers we were privileged to witness all 100 of these rare species with which we share this short time and small place in the universe, and to revel in the richness of their creation. Connecting with these creatures made us vividly aware of how much the natural world means to us. The experience of kinship with what we were beholding evoked in us a feeling of tenderness toward these complex and beautiful lives embedded in exquisitely evolved domains. Our hope is that the splendor of these creatures shines through in these photographs, and that they reveal themselves to you as they have to us.

Inside Madison Saltpetre Cave photographing the Madison Cave isopod

Introduction

EDWARD O. WILSON

LOOK AT THE FRESHNESS AND VIGOR of the organisms in these portraits and consider: no species dies of old age. Every species that disappears is killed and it dies young, at least in a physiological sense. We still occasionally hear someone call the California condor a senescent species whose time has come. Don't hold on too tight, the prescription follows, let it go! That opinion is based on a false analogy with organisms, which compares an endangered species to a terminal patient in intensive care too expensive for society to prolong. The truth is that the great majority of such species are composed largely of young, healthy individuals, just like other, more fortunate species that are still widespread. The condor disappeared from the wild not because its heredity declined but because people destroyed most of its natural habitat and shot and poisoned the dwindling remnant. When only a dozen individuals remained in the wild, they were captured and placed with a confined breeding colony near San Diego. Given protection and food, they and their offspring are now flourishing. If the condor habitat were somehow restored across the prehistoric breeding range and the species left alone within it for a few decades, *Gymnogyps californianus* would return as an abundant bird across hundreds of thousands of square miles of the American landscape.

The condor didn't change; we did. Extremely few species are dying for any reason on their own, because humanity makes the decisions of life and death at this point in geological time. Paleontologists estimate from the fossil record that before the origin of *Homo sapiens,* each species and its immediate descendants lived an average of between 1 and 10 million years: 6 million years in the case of starfish and other echinoderms, 1 to 10 million years for flowering plants, .5 to 5 million years for mammals, and so on. Through most of geological history, outside rare catastrophic disturbances of the climate, the probability of extinction has also remained more or less constant through time. Consider the following oversimplified example of such a trajectory. If one-half the species of a certain group were alive at the end of a million years, then one-half of those (or one-quarter of the original) persisted 2 million years, and half of those again (one-eighth of the original) lasted 3 million years, and so on until all or almost all were gone. The overall rate of extinction, again according to group, has averaged from one species per million to one species per 10 million each year. In the present century human activity has multiplied this rate between 1,000 and 10,000 times. Furthermore, the toll is steadily accelerating. The losses caused during the next 30 years, not including the impact of earlier declines, could easily reach 20 percent of the plant and animal species of the world.

By far the most important agent of extinction is habitat destruction. Bear with me while I cite several more figures to frame the magnitude of this phenomenon. All biologists know that when habitat is reduced, species disappear. The rate of loss, established by hundreds of independent studies on many kinds of plants and animals, has been found in the great majority of cases to fall between the third and sixth root of the area. A common intermediate value encountered in these analyses is the fourth root. The fourth root of the habitat area translates to the following rule of thumb: a 90 percent loss of area results in an eventual 50 percent loss of species. The rate of loss of tropical rain forest during the late 1980s has been estimated to have been about 1.8 percent per year, having nearly doubled since the 1970s. That amount translates, at the "typical" value of the fourth root, to half a percent of the species lost or doomed annually. At the theoretically lowest, least destructive value, it still projects to a quarter percent extinction per year.

As the forest is cut back, as dams fill valleys and obliterate natural stream systems, as wild savanna is ploughed under, some very localized species are extinguished at once. Others persist in populations too small to persist beyond a few more years. The ivory-billed woodpecker survived for a while in the diminished old-growth forests of the southern United States, but finally went extinct despite efforts toward the end to save it. A related form hangs on in a remote forest of Cuba's Oriente Province, its long-term future grim. Another close relative, the giant imperial woodpecker, also dwindled when the old-growth pine forests of the Mexican Plateau were cut. One of the last known living individuals was shot for food in the 1960s by a local truck driver, who when later interviewed (by George Plimpton) rated it "a great piece of meat."

Plant species pushed to the edge of extinction by habitat destruction provide a set of equally dispiriting dramas. In tiny forest reserves of western Ecuador, comprising the last remnants left after widespread clear-cutting, are tree species known from only one or several individuals, too few to reproduce naturally. They are, to use the black-humor notation of conservation biologists, the "living dead." By 1986 *Banara vanderbiltii*, a small tree of Puerto Rico's moist limestone forest, had been reduced to two plants growing on a farm near Bayamon. At the eleventh hour, botanists obtained cuttings and are now successfully growing them in the Fairchild Tropical Garden in Miami. The sole living individual of the Hawaiian shrub *Kokia cookei*, whose portrait is included in the present volume, survives as a graft on the trunk of another species of *Kokia*.

How long do doomed species last in habitats fragmented too small to support them? It depends on the size of the reserve. The loss of bird species locally in tropical American rain forest patches reduced to 1 to 25 square kilometers has been observed to reach 10 percent to 50 percent in the first 100 years. Animals that are sparsely distributed, specialized in their requirements, and short-lived, such as certain birds and butterflies, disappear first. Tree species can last for decades or in extreme cases for centuries. But a severe storm, a fire, or an errant chain saw can finish any of them in an instant.

The life of endangered species can be extended indefinitely by propagation in zoological parks and botanical gardens. Yet even when successful, it should give little comfort. Only a tiny fraction can be saved in such a manner. Facilities exist for up to several tens of thousands of favored higher plants and vertebrates, but for most plants and vertebrates and, worse, for almost all the millions of insects, other invertebrates, fungi, and bacteria so vital to the functioning of the natural world, there is no such refuge.

As habitat is reduced, the other three apocalyptic horsemen of extinction—pollution, overharvesting, and the introduction of exotic species—come into play. The worst is the last of these. Under certain conditions a small number of exotic species can alter entire ecosystems and diminish or extinguish indigenous species. Some of my favorite organisms, the ants, are especially dangerous. The African species *Pheidole megacephala*, accidentally introduced into Hawaii in the 1800s by European commerce, proceeded to wipe out a large fraction of the indigenous insect faunas. Because there were no native ants in the islands, the resident insects were vulnerable to any such highly organized invader. Many were flightless and not equipped with the stings and chemical repellants needed to escape from the relentless swarms of the *Pheidole*, which spread into virtually every leaf sprig and soil crevice. One entomologist observing the early stages of the invasion in the 1890s watched as forest trees were swept clean of the old fauna.

P. megacephala is an example of a keystone species, any plant or animal whose presence—or removal—has a widespread effect on the remainder of the ecosystem. An increasing number of such species are coming to light as ecological studies deepen. Within the United States, they have proved most ruinous in Hawaii, California, and southern Florida, but they can be found in other states and in a variety of habitats. One of many is *Orconectes rusticus*, an exotic crayfish established in the lakes of northern Wisconsin. It has driven native crayfish to local extinction by competition and its own greater resistance to predators. Its dense populations have cropped other local invertebrates and water plants to low levels or even extinction. The alteration is expected to reduce fish populations dependent on these food sources, and to diminish the productivity of the lakes as a whole.

Local mass extinctions are being observed with increasing frequency around the world in groups as disparate as flowering plants and freshwater fishes. They often entail the extirpation of species and races found nowhere else. It is a sad rule of field biology that when ecosystems are studied carefully before and after serious human disturbance, extinctions are almost always revealed. The casualty roll includes the following.

- One-fifth of all the bird species of the world during the past 2,000 years, with 11 percent of the remaining 9,040 species currently threatened or endangered.

- More than half the 266 species of exclusively freshwater fishes in peninsular Malaysia.

- Fifteen of the 18 endemic freshwater fishes of Lake Lanao in the Philippines.

- All of the 11 endemic tree snail species of Moorea in the Society Islands. Those on nearby Tahiti, as well as on Hawaii, are also rapidly disappearing.

- Upwards of 90 endemic or near-endemic plant species from a single mountain ridge in Ecuador, through clear-cutting of forest during 1978–86.

During the past 100 years the United States has suffered extinction of a considerable number of its endemic species: 2.3 percent of the birds, 2.2 percent of the amphibians, 1.2 percent of the freshwater fishes, 1.1 percent of the plant species, and a staggering 8.6 percent of the freshwater mussels.

Most of the losses just listed, including all of those from the United States, are of full species over all their ranges. They are, to use the parlance of the conservation biologists, global and not just local. The toll is far higher, and still largely unmeasured, for geographical races (subspecies, as biologists call them), which are populations that possess distinctive traits and occupy separate geographic ranges, but still belong to the same species as other races classified with them. The Cuban and newly extinct North American ivory-billed woodpeckers, for example, are regarded as subspecies of the same species, *Campephilus principalis*. The closely related imperial woodpecker of Mexico, on the other hand, is different enough in size and plumage to be classified as a full species, *C. imperialis*. Ornithologists guess that if the two forms of ivory-bills had occurred together, they would have interbred freely, but the imperial would have remained separate from both.

Finally, as populations are restricted in size, they lose some of the genetic varieties composing the internal variability of the species. When individual organisms become more uniform within the populations, the species as a whole loses its ability to adapt to changes in the environment. The organisms also become in effect more closely related to one another (because their genes are overall more similar). The rate of inbreeding consequently increases, which in a few extreme cases reduces vigor and fertility of individuals and renders the species as a whole still more vulnerable.

Not just species and genetic variants but entire ecosystems vanish as the natural environment is eroded. A final stand of old-growth forest leveled here or a river dammed there can erase large numbers of species in one stroke. When the last remnant of a rain forest tract is clear-cut in the Andean foothills, to take one extreme example, we can expect to see the local demise of perhaps 300 kinds of birds, up to 1,000 butterflies, 300 ants, 50,000 beetles, 1,000 trees, tens of thousands of fungi and bacteria, and so on down a long roster of major groups. Some of the species can be expected to have been limited to that particular habitat, so their extinction will be total and final.

Each species in the ecosystem occupies a precise niche, demanding a certain location in the forest, an exact microclimate, and particular nutrients and temperature and humidity cycles with specified timing to trigger phases of the life cycle. Many and perhaps most of the species are locked in symbioses with other species. They are the pollinators, root symbionts, epiphytes, and other mutualistic partners, hangers-on, and parasites that raise the superstructure to extreme complexity. They cannot survive and reproduce unless joined with certain other species in the correct configurations of space and time.

Each species, to put the matter succinctly, is a masterpiece. It deserves that rank in the fullest sense: a creation assembled with extreme care by genius. The master craftsman that shaped it was natural selection, acting upon mutations and recombinations of genes, through a vast number of steps over a long period of time. Each species is consequently a source of scientific knowledge and aesthetic pleasure. The number of genes prescribing an organism belonging to a higher life-form, in other words more complex than a bacterium, runs into the hundreds of thousands; the nucleotide pairs composing them, the genetic letters that encode the life-giving enzymes, number according to species from 1 billion to 10 billion. If the DNA helices in one cell of a mouse (a typical animal species) were placed end on end and magically enlarged to the same width as a piece of wrapping string, they would extend for about 600 miles, with 20-odd nucleotide pairs packed into every inch. How all that genetic information translates into a fully functioning organism is still partly a mystery. The lesson to be drawn is that the life-forms around us are too old, too complex, and too valuable to be carelessly discarded.

So far I have offered some of the facts and statistics from which scientists argue the case for conservation. Such information has on occasion changed thinking at a Senate hearing or a conference of the National Research Council. It implies cost and benefit, the national interest, the state of the world. Appealing to a pragmatic worldview and criteria already accepted, it is an argument primarily of the mind. "Believe me," the scientist says, "the evidence is persuasive: a real problem exists, and it is worthy of your serious attention."

The remarkable portraits presented herein by Susan Middleton and David Liittschwager have a wholly different impact: they speak to the heart. In the end their kind of testimony may count as much toward conserving life as all the data and generalizations of science. Each photograph strips the environment from the background of the animal or plant chosen to represent its species, forcing us to look at the face of the organism unadorned. We are given no opportunity for distraction. The technical quality of the portraiture is so fine, the composition so pleasing, that we gladly contemplate for a few moments creatures as strange as the Furbish lousewort (yes, at last we have a look at the plant with the fabled name) and the American burying beetle. Each species is personalized. The freshness of the visage of the emissary chosen by Middleton and Liittschwager, the sign of a life cycle still capable of renewal, offers hope. The creature can no longer be called a weed, a flower, or a bug. It has a name, a million-year history, and a place in the world. It is now unmistakable and unavoidable. It is more than a statistic, reduced to say one less warbler from the forests of North America, one less bivalve from the riffles of the Tennessee River drainage. The endangered species whose portrait we see can be a companion henceforth, if we wish, for us and for our descendants as far into the future as can be imagined. Its death we will have occasion to mourn bitterly.

California Condor
Gymnogyps californianus

STATUS	DATE LISTED	POPULATION		HABITAT	RANGE	THREATS
Endangered	March 11, 1967	76		Caves and crevices on rocky cliffs for nesting and roosting, open grassland and oak-savanna for foraging	Sespe Condor Sanctuary, Los Padres National Forest, California	Shooting, poisoning, collision with power lines, residential development

PHOTOGRAPHED
September 18–23, 1993
Los Angeles Zoo
Los Angeles, California

21

Red Wolf
Canis rufus

STATUS	DATE LISTED	POPULATION	HABITAT	RANGE	THREATS
Endangered	October 3, 1970	Indeterminate	Dense mountain and bottomland forests, coastal prairies, marshes	Central Texas east to coast of Florida and Georgia; Mississippi River Valley north to central Illinois and Indiana	Low numbers, hunting, habitat loss, interbreeding

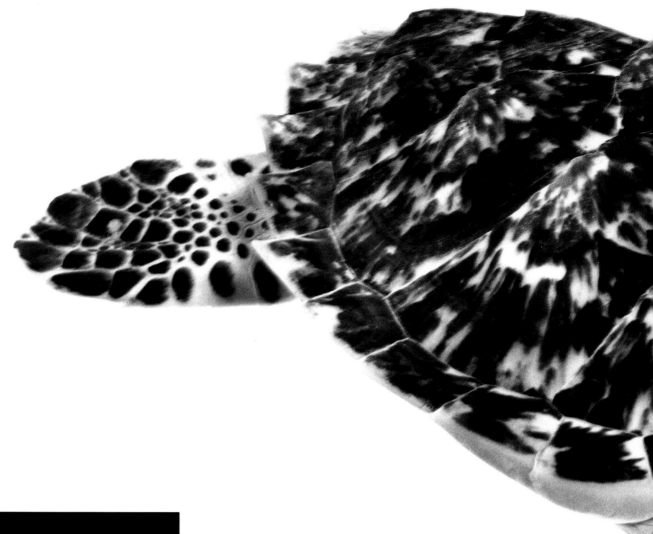

PHOTOGRAPHED
March 23, 1992
Aquarium of the Americas
New Orleans, Louisiana

STATUS	DATE LISTED	POPULATION	HABITAT	RANGE	THREATS
Endangered	June 2, 1970	Indeterminate	Tropical ocean water, isolated, undisturbed beaches for nesting	Gulf of Mexico, tropical Atlantic Ocean	Shrimpers' nets, egg poaching, tortoiseshell trade, real estate development

Hawksbill Sea Turtle

Eretmochelys imbricata

Presidio Manzanita

Arctostaphylos pungens var. *ravenii*

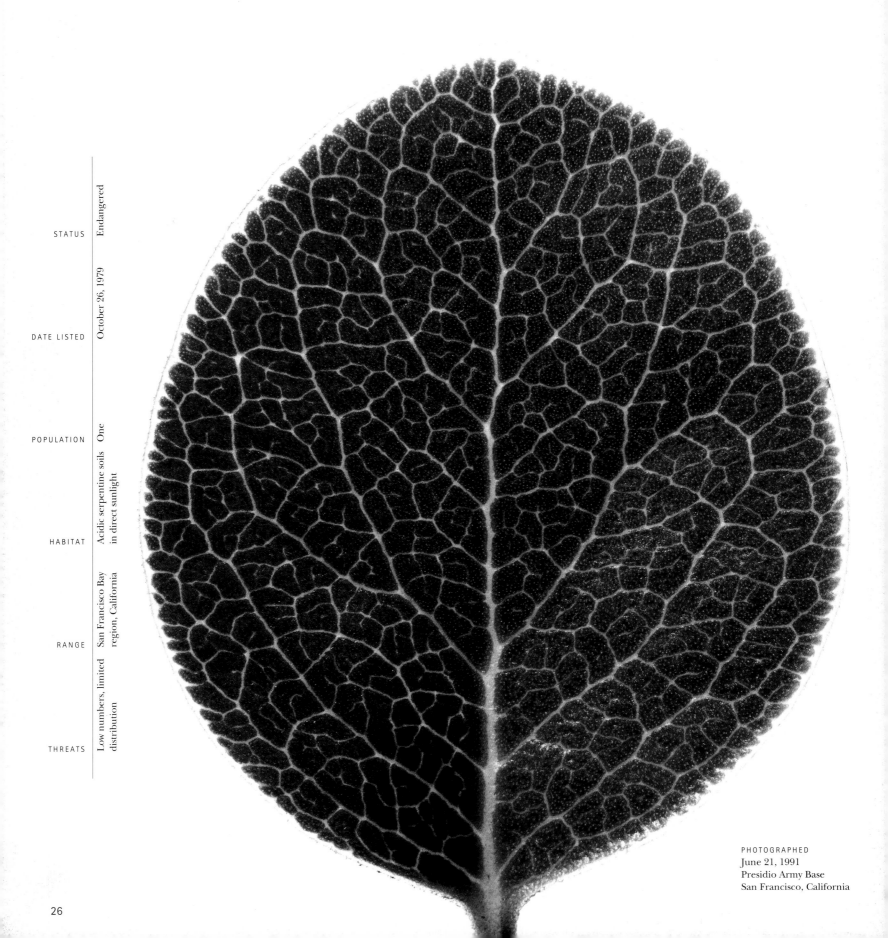

STATUS	Endangered
DATE LISTED	October 26, 1979
POPULATION	One
HABITAT	Acidic serpentine soils in direct sunlight
RANGE	San Francisco Bay region, California
THREATS	Low numbers, limited distribution

PHOTOGRAPHED
June 21, 1991
Presidio Army Base
San Francisco, California

PHOTOGRAPHED
November 18, 1991
Sybille Wildlife Research and
Conservation Education Center
Wheatland, Wyoming

Wyoming Toad

Bufo hemiophrys baxteri

STATUS	DATE LISTED	POPULATION	HABITAT	RANGE	THREATS
Endangered	January 17, 1984	50	Wetlands	Laramie Basin, southeast Wyoming	Possibly pesticides, predation, irrigation drainage practices

Boulder Darter

Etheostoma wapiti

PHOTOGRAPHED
October 30, 1992
National Fisheries Research Center
Gainesville, Florida

STATUS	DATE LISTED	POPULATION	HABITAT	RANGE	THREATS
Endangered	September 1, 1988	Eight known populations	Boulders in river runs of moderate depth	Elk River in Tennessee and Alabama and two tributaries	Dam construction, water pollution, sedimentation

Missouri Bladderpod

Lesquerella filiformis

STATUS	Endangered
DATE LISTED	February 9, 1987
POPULATION	Indeterminate
HABITAT	Shallow soil in limestone glades
RANGE	Southwest Missouri
THREATS	Overgrazing by livestock, urbanization, competition from exotic plants, suppression of natural fires

PHOTOGRAPHED
May 7, 1992
Wilson's Creek National Battlefield
Republic, Missouri

Swamp Pink

Helonias bullata

PHOTOGRAPHED
May 13, 1992
New York Botanical Garden
Bronx, New York

PHOTOGRAPHED
May 22, 1992
Garden in the Woods
Framingham
Massachusetts

STATUS	DATE LISTED	POPULATION	HABITAT	RANGE	THREATS
Threatened	September 9, 1988	120 known sites	Freshwater wetlands	Atlantic seaboard from southern New York to South Carolina	Filling, draining, and degradation of wetlands

PHOTOGRAPHED
March 29, 1993
University of Florida
Gainesville, Florida

Schaus Swallowtail Butterfly

Heraclides aristodemus ponceanus

STATUS	DATE LISTED	POPULATION	HABITAT	RANGE	THREATS
Endangered	August 31, 1984	Fewer than 100	Hardwood hammock	Key Biscayne National Park, Florida	Urban development, pesticide poisoning, storms

Aleutian Canada Goose

Branta canadensis leucopareia

STATUS	DATE LISTED	POPULATION	HABITAT	RANGE	THREATS
Threatened	March 11, 1967	8,000	Offshore islands, cropland, pasture, marshes	Six of the Aleutian Islands and two of the Samedi Islands in Alaska, Oregon, California, British Columbia	Predation by foxes

PHOTOGRAPHED
December 10, 1991
Humbolt State University
Arcata, California

PHOTOGRAPHED
May 9, 1992
Missouri Botanical Garden
St. Louis, Missouri

Running Buffalo Clover
Trifolium stoloniferum (Fabacae)

STATUS	DATE LISTED	POPULATION	HABITAT	RANGE	THREATS
Endangered	June 5, 1987	Indeterminate	Dry upland forests, floodplain forests, streambed gravels	West Virginia, Indiana, Ohio, Kentucky	Loss of large herbivores that open up the plant's habitat

Fat Pocketbook Pearly Mussel

Potamilus capax

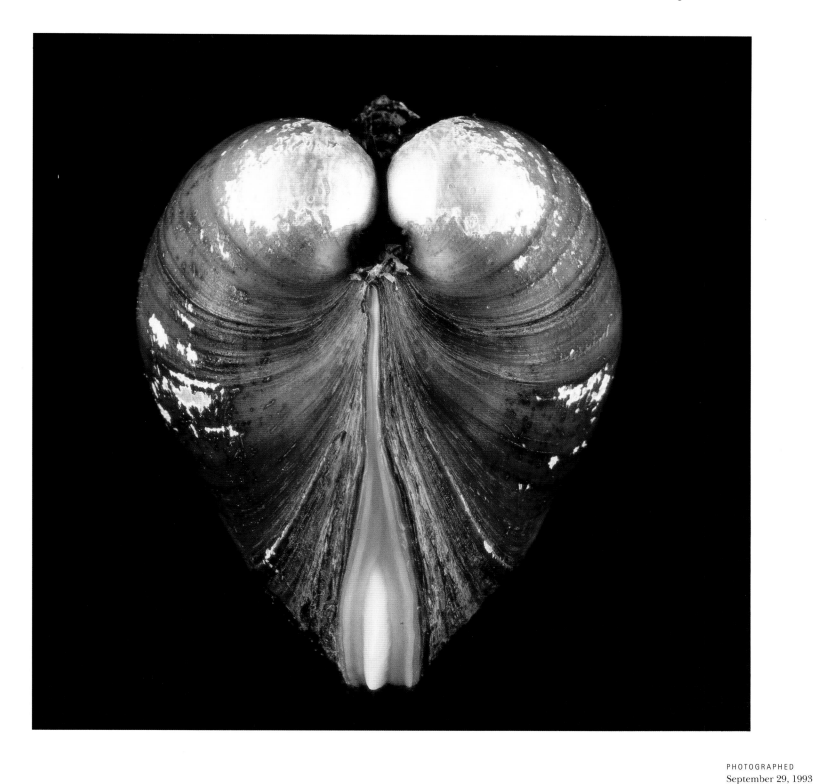

PHOTOGRAPHED
September 29, 1993
St. Francis River
Madison, Arkansas

STATUS	DATE LISTED	POPULATION	HABITAT	RANGE	THREATS
Endangered	June 14, 1976	Nine known populations of indeterminate number	Slow-moving water with sandy, muddy substrate	St. Francis River and St. Francis Floodway, Arkansas; Wabash River, Indiana	Damming, dredging, agricultural chemicals

PHOTOGRAPHED
November 19, 1992
Cincinnati Zoo
and Botanical Garden
Cincinnati, Ohio

Jaguarundi

Felis yagouaroundi

STATUS	DATE LISTED	POPULATION	HABITAT	RANGE	THREATS
Endangered	June 14, 1976	Indeterminate	Chapparal, mesquite thickets near streams	Possibly Texas and southern Arizona, Mexico through Central America to Panama	Loss of riparian brush-land to agriculture and grazing pastures

Plymouth Redbelly Turtle
Pseudemys rubiventris bangsii

PHOTOGRAPHED
June 19, 1992
Rutland, Massachusetts

STATUS	DATE LISTED	POPULATION	HABITAT	RANGE	THREATS
Endangered	April 2, 1980	300	Ponds and pond banks	Plymouth County, Massachusetts	Real estate development, predation

STATUS	DATE LISTED	POPULATION	HABITAT	RANGE	THREATS
Endangered	June 2, 1979	Indeterminate	Conifer and mixed conifer-deciduous forests, 6,000–9,000 feet	Arizona, northern and central Mexico	Illegal collecting, destruction and fragmentation of forest habitat

Thick-Billed Parrot

Rhynchopsitta pachyrhyncha

PHOTOGRAPHED
August 15, 1991
Sacramento Zoo
Sacramento, California

Echinocereus chisoensis var. *chisoensis (=reichenbachii)*

Chisos Mountain Hedgehog Cactus

PHOTOGRAPHED
July 7, 1993
Chihuahuan Desert Visitor Center
near Fort Davis, Texas

STATUS	DATE LISTED	POPULATION	HABITAT	RANGE	THREATS
Threatened	September 30, 1992	11 known populations totaling 100–200 individuals	Alluvial flats	East of Chisos Mountains, in Big Bend National Park, Texas	Illegal collecting, limited distribution

Nellie Cory Cactus

Coryphantha minima

PHOTOGRAPHED
April 22, 1993
Desert Botanical Garden
Phoenix, Arizona

STATUS	DATE LISTED	POPULATION	HABITAT	RANGE	THREATS
Endangered	November 7, 1979	Three known sites	High desert scrub on gravelly soils	Chihuahuan Desert, Brewster County, Texas	Commercial collecting

Large-Fruited Sand Verbena

Abronia macrocarpa

STATUS	DATE LISTED	POPULATION	HABITAT	RANGE	THREATS
Endangered	September 28, 1988	Three known sites totaling a few thousand individuals	Stabilized sandy areas	Sand dunes of central-eastern Texas	Off-road vehicles, real estate development

PHOTOGRAPHED
April 30, 1993
Southwestern Texas State University
Botany Department
San Marcos, Texas

PHOTOGRAPHED
August 15, 1992
Elkins, West Virginia

Triodopsis platysayoides

Flat-Spired Three-Toothed Land Snail

STATUS	DATE LISTED	POPULATION	HABITAT	RANGE	THREATS
Threatened	July 3, 1978	18 known populations, most with fewer than 10 adults	Deep leaf litter in sandstone-based, mixed forests	Cheat River Canyon, West Virginia	Limited distribution, trampling by hikers, forest fires

21 days after spawning

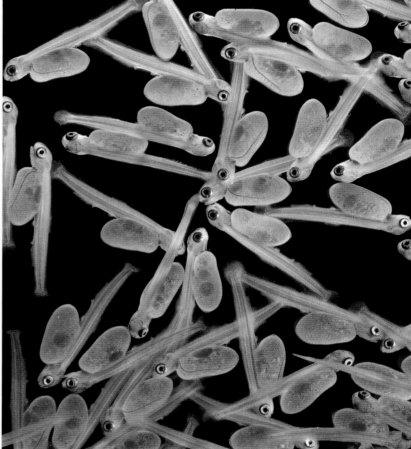

45 days after spawning

75 days after spawning

Winter-Run Chinook Salmon

Oncorhynchus tschawytscha

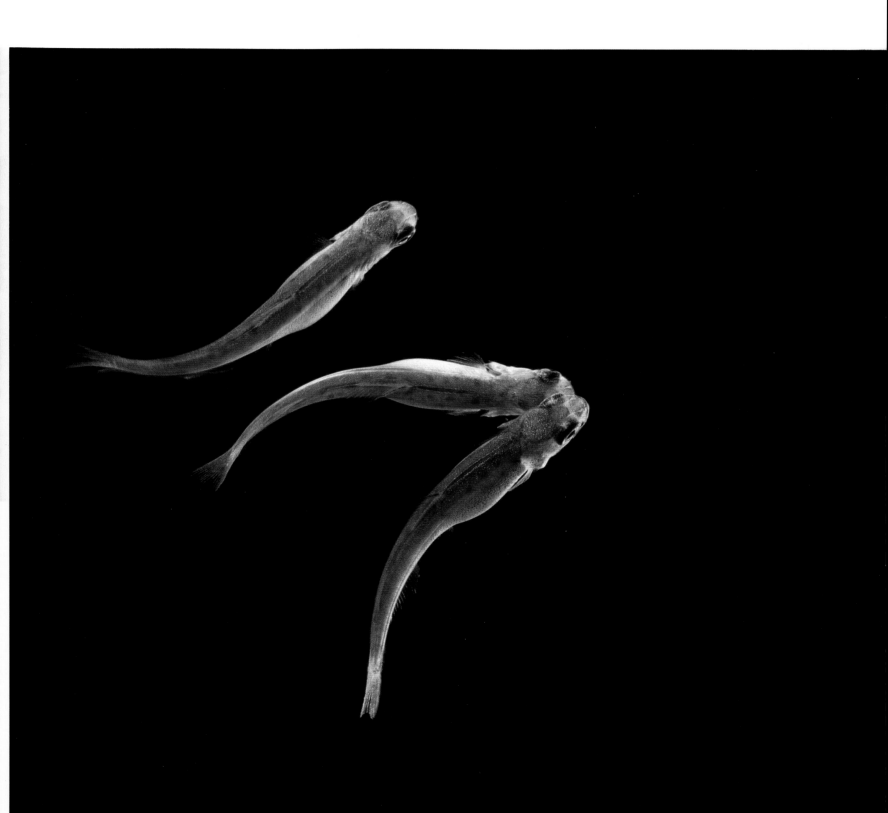

STATUS	DATE LISTED	POPULATION	HABITAT	RANGE	THREATS
Threatened	August 4, 1989	191 (1991 run), 1,180 (1992 run)	Oceans and gravel-bottomed freshwater rivers	Eastern Pacific Ocean from central California to the Washington border, Sacramento River for spawning	Dams and water diversion, water pollution, predation by native fish

PHOTOGRAPHED
July 20, 1993
Coleman National Fish Hatchery
Anderson, California

Florida Panther
Felis concolor coryi

PHOTOGRAPHED
October 7, 1992
Lowry Park Zoological Garden
Tampa, Florida

four months old

PHOTOGRAPHED
March 21, 1992
White Oak Conservation Center
Jacksonville, Florida

STATUS	DATE LISTED	POPULATION	HABITAT	RANGE	THREATS
Endangered	March 11, 1967	30–50	Pine and hammock forests, wetlands	South Florida	Industrial, urban, and agricultural expansion, hybridization with non-native pumas

Steller's Sea Lion

Eumetopias jubatus

PHOTOGRAPHED
November 10, 1992
Mystic Marinelife Aquarium
Mystic, Connecticut

STATUS	DATE LISTED	POPULATION	HABITAT	RANGE	THREATS
Threatened	April 4, 1990	40,000	Open ocean, low rocky outcrops	North Pacific Ocean from central California to Japan	Commercial fishing: competition for fish, incidental drowning in nets, deliberate killing by fishermen

Pitcher's Thistle

Cirsium pitcheri

STATUS	DATE LISTED	POPULATION	HABITAT	RANGE	THREATS
Threatened	July 18, 1988	143 known sites	Lakeshore sand dunes	Michigan, Wisconsin, Indiana; Ontario, Canada	Real estate development, recreational use of habitat, sand mining, dune stabilization

PHOTOGRAPHED
June 25, 1993
Holden Arboretum
Kirtland, Ohio

PHOTOGRAPHED
July 2, 1993
Holden Arboretum
Kirtland, Ohio

New Mexico Ridgenose Rattlesnake

Crotalus willardi obscurus

PHOTOGRAPHED
December 18, 1992
Museum of Vertebrate Zoology
University of California
Berkeley, California

STATUS	DATE LISTED	POPULATION	HABITAT	RANGE	THREATS
Threatened	August 4, 1978	Indeterminate	Pine-oak woodland in narrow mountain canyons	Animas and Peloncillo mountains of south-western New Mexico and the Sierra San Luis in northern Chihuahua and Sonora, Mexico	Low numbers, restricted range, illegal collecting

Gulf Sturgeon

Acipenser oxyrhynchus desotoi

STATUS	DATE LISTED	POPULATION	HABITAT	RANGE	THREATS
Threatened	September 30, 1991	Indeterminate	Clear, cool rivers with sandy bottoms	Eastern Gulf of Mexico and adjacent streams	Dammed rivers, poaching

PHOTOGRAPHED
March 25, 1992
Aquarium of the Americas
New Orleans, Louisiana

Dwarf Wedge Mussel

Alasmidonta heterodon

PHOTOGRAPHED
June 21, 1992
Sumner Falls
Connecticut River, Vermont

STATUS	DATE LISTED	POPULATION	HABITAT	RANGE	THREATS
Endangered	March 14, 1990	19 known populations	Mud, sand, gravel stream bottoms	Atlantic Coast river systems	Riverside development, dam construction and water projects, water pollution, cattle grazing

STATUS	Endangered
DATE LISTED	July 22, 1988
POPULATION	Indeterminate
HABITAT	Clear, slow-moving freshwater streams
RANGE	Headwaters streams of James River, Virginia
THREATS	Agricultural runoff, stream channelization, competition with the Asian clam

James Spinymussel
Pleurobema collina

PHOTOGRAPHED
August 10, 1992
South bank of Saint John River
Allagash, Maine

Furbish Lousewort

Pedicularis furbishiae

STATUS	DATE LISTED	POPULATION	HABITAT	RANGE	THREATS
Endangered	April 26, 1978	5,000–18,000	Riparian habitat, mostly periodically disturbed, north-facing, well-drained sandy loam	Saint John River Valley in northern Maine and western New Brunswick, Canada	Proposed hydroelectric projects, agricultural and residential development, logging, low genetic variation

Black-Footed Ferret
Mustela nigripes

PHOTOGRAPHED
November 17, 1991
Sybille Wildlife Research
and Conservation
Education Center
Wheatland, Wyoming

STATUS	DATE LISTED	POPULATION	HABITAT	RANGE	THREATS
Endangered, on the verge of extinction	June 2, 1970	435	Prairie dog grassland communities	Laramie Basin, southeast Wyoming	Destruction of prairie dogs and grasslands

Florida Scrub Jay

Aphelocoma coerulescens coerulescens

STATUS	Threatened
DATE LISTED	June 3, 1987
POPULATION	7,000–11,000
HABITAT	Dense thickets of scrub oaks amid ancient sand dunes
RANGE	Florida
THREATS	Real estate development, orange groves, suppression of natural fires, lack of prescribed burning

PHOTOGRAPHED
February 27, 1992
Archbold Biological Station
Lake Placid, Florida

77

Florida Torreya

Torreya taxifolia

STATUS	Endangered
DATE LISTED	January 23, 1984
POPULATION	1,000
HABITAT	Shaded bluffs and ravines
RANGE	Apalachicola River Valley of the Florida panhandle and southern Georgia
THREATS	Fungal disease, low numbers, suppression of natural fires

PHOTOGRAPHED
August 24, 1992
Biltmore Estate
Asheville, North Carolina

PHOTOGRAPHED
June 25, 1992
Brooklyn Botanical Garden
Brooklyn, New York

PHOTOGRAPHED
October 28, 1992
National Fisheries Research Center
Gainesville, Florida

STATUS	DATE LISTED	POPULATION	HABITAT	RANGE	THREATS
Endangered	March 11, 1967	Indeterminate	Deep pools in estuaries, tidal and freshwater rivers and streams, occasionally open ocean	Atlantic Coast from New Brunswick, Canada, to Florida	Dams, industrial pollution, sedimentation

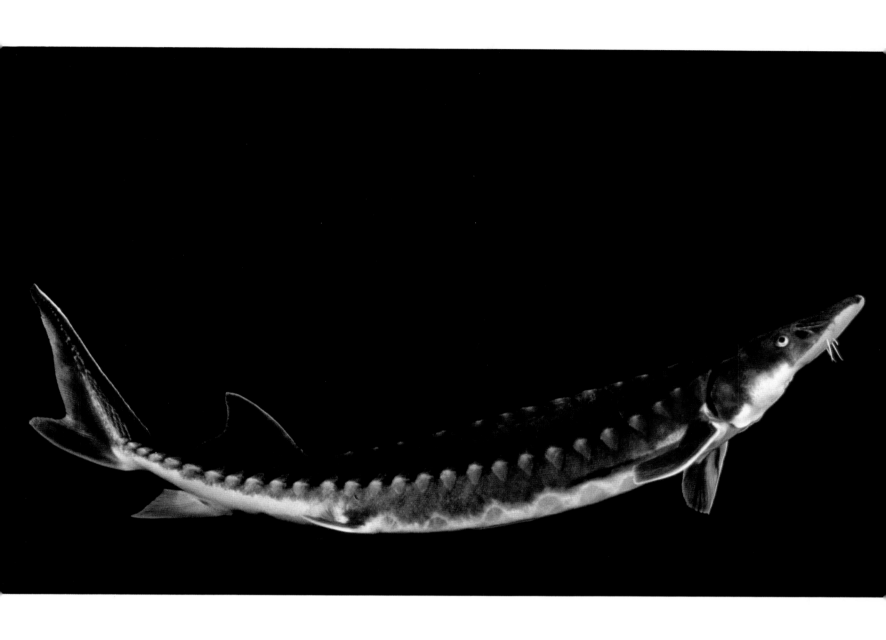

Shortnose Sturgeon

Acipenser brevirostrum

PHOTOGRAPHED
October 8, 1992
Busch Gardens Zoo
Tampa, Florida

American Crocodile

Crocodylus acutus

STATUS	DATE LISTED	POPULATION	HABITAT	RANGE	THREATS
Endangered	December 18, 1979	200–300 in the U.S.	Mangrove-lined saltwater estuaries	Southern tip of Florida, West Indies, east coast of Central America	Habitat loss, nest destruction, road-kills, predation

Arizona Agave

Agave arizonica

STATUS	DATE LISTED	POPULATION	HABITAT	RANGE	THREATS
Endangered	May 18, 1984	60 clones	Stony drainage bottoms to rocky mountain saddles, 3,000–6,000 feet	Central mountains of Arizona	Grazing cattle and deer

PHOTOGRAPHED
February 4, 1992
Desert Botanical Garden
Phoenix, Arizona

Northern Swift Fox

Vulpes velox hebes

STATUS	Endangered
DATE LISTED	June 2, 1970
POPULATION	Indeterminate
HABITAT	Short- and midgrass prairies
RANGE	South Dakota, Colorado, Wyoming, Alberta, Saskatchewan, and Manitoba, Canada
THREATS	Trapping, poisoning, conversion of grassland to farmland

PHOTOGRAPHED
November 24, 1992
Cochrane Wildlife Reserve
Cochrane, Alberta, Canada

Kemp's Ridley Sea Turtle

Lepidochelys kempii

STATUS	DATE LISTED	POPULATION	HABITAT	RANGE	THREATS
Endangered	December 2, 1970	Fewer than 700 mature females, indeterminate number of males and subadults	Coastal waters of warm oceans, sandy beaches for nesting	Primarily throughout Gulf of Mexico, also along Atlantic Coast from Bermuda to Nova Scotia and across to the Azores; nests on Rancho Nuevo Beach, Tamaulipas, Mexico	Drowning in trawling gear; pollution, particularly ingesting plastic

PHOTOGRAPHED
May 13, 1993
National Marine Fisheries Service
Galveston, Texas

Piping Plover
Charadrius melodus

PHOTOGRAPHED
June 8, 1993
Third Cliff
Scituate, Massachusetts

STATUS	DATE LISTED	POPULATION	HABITAT	RANGE	THREATS
Threatened on Atlantic Coast and Great Plains, endangered in Great Lakes area	December 11, 1985	2,400 pairs in U.S. and Canada	Sand or gravel beaches, sandbars	Atlantic Coast, Great Lakes, Great Plains	Residential development, water projects, off-road vehicles, predation by mammals, human disturbance

Oenothera avita var. *eurekensis*

Eureka Valley Evening Primrose

PHOTOGRAPHED
May 25, 1993
Eureka Valley, California

STATUS	DATE LISTED	POPULATION	HABITAT	RANGE	THREATS
Endangered	April 26, 1978	Indeterminate	Shallow, windblown sand on the lower slopes and base of dunes	Eureka Valley, northern Mojave Desert, California	Competition from non-native plants, off-road vehicles

PHOTOGRAPHED
April 27, 1993
San Antonio Zoological
Gardens and Aquarium
San Antonio, Texas

Whooping Crane

Grus americana

STATUS	DATE LISTED	POPULATION	HABITAT	RANGE	THREATS
Endangered	March 11, 1967	174 in the wild, 117 in captivity	Wilderness wetlands	Alberta and Northwest Territories, Canada, and Wyoming and Idaho in summer; Texas and New Mexico in winter	Destruction of wetlands, collision with power lines, avian tuberculosis

Harperella

Ptilimnium nodosum

STATUS	Endangered
DATE LISTED	September 28, 1988
POPULATION	16 known populations
HABITAT	Shallow, oft-flooded coastal-plain ponds or rocky beds along clear, rushing streams
RANGE	Alabama, Arkansas, Georgia, the Carolinas, West Virginia, Maryland
THREATS	Filling of wetlands, changes in water level, residential development

PHOTOGRAPHED
August 14, 1992
Sideling Hill Creek
Hancock, Maryland

Sternotherus depressus
Flattened Musk Turtle

PHOTOGRAPHED
November 17, 1992
Columbus Zoological Gardens
Powell, Ohio

STATUS	DATE LISTED	POPULATION	HABITAT	RANGE	THREATS
Threatened	June 11, 1987	Indeterminate	Free-flowing rivers and creeks, freshwater pools	Black Warrior River system, Alabama	Dam construction, water pollution from mining, illegal collecting, hybridization

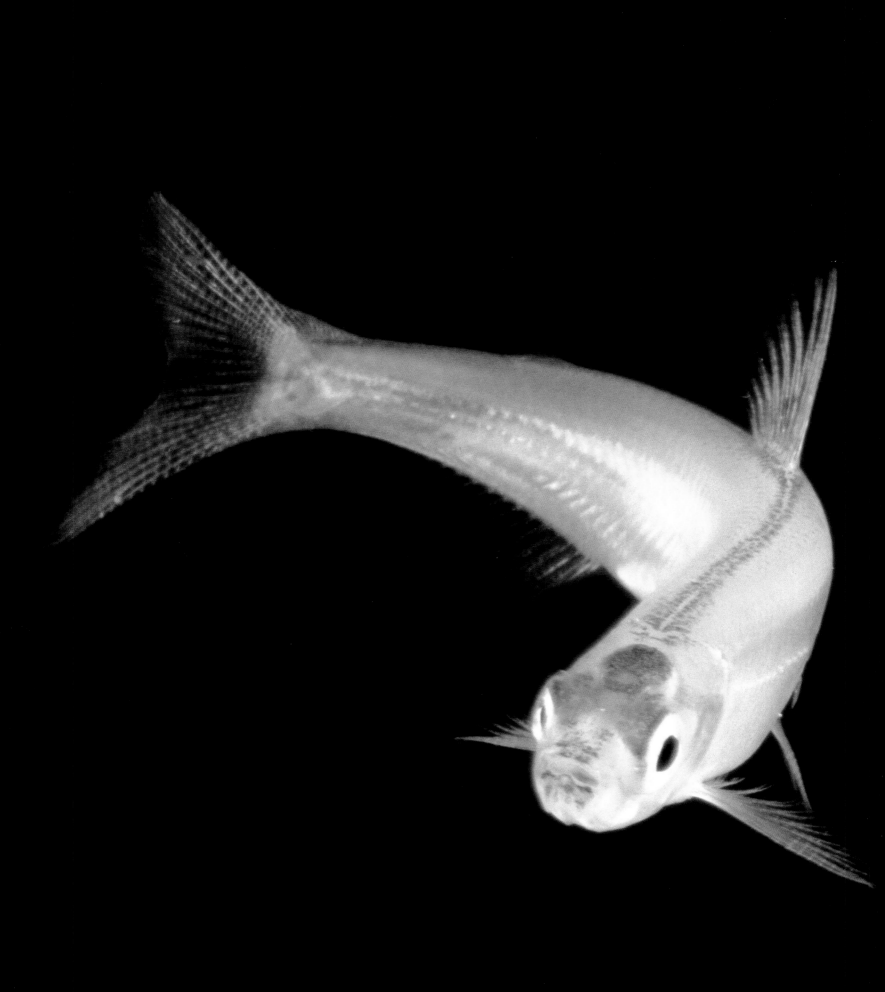

Delta Smelt

Hypomesus transpacificus

STATUS	**DATE LISTED**	**POPULATION**	**HABITAT**	**RANGE**	**THREATS**
Threatened	March 5, 1993	Indeterminate	Low-salinity surface and shallow estuary waters	Upper estuary of the Sacramento and San Joaquin rivers, California	Freshwater diversions for agricultural and urban use, prey decline due to introduced species, pesticides

PHOTOGRAPHED
December 13, 1991
Pittsburg /Antioch, California

PHOTOGRAPHED
June 23, 1993
Colyer Prairie
near Lawrence, Kansas

STATUS	DATE LISTED	POPULATION	HABITAT	RANGE	THREATS
Threatened	September 1, 1988	Indeterminate	Deep-loamed, unplowed prairie, igneous glades	Illinois, Iowa, Kansas, Missouri	Urban and agricultural development

Mead's Milkweed

Asclepias meadii

Blue-Tailed Mole Skink

Eumeces egregius lividus

STATUS	DATE LISTED	POPULATION	HABITAT	RANGE	THREATS
Threatened	November 6, 1987	Indeterminate	Sand pine scrub	Central Florida	Suppression of natural fires, agricultural and residential development

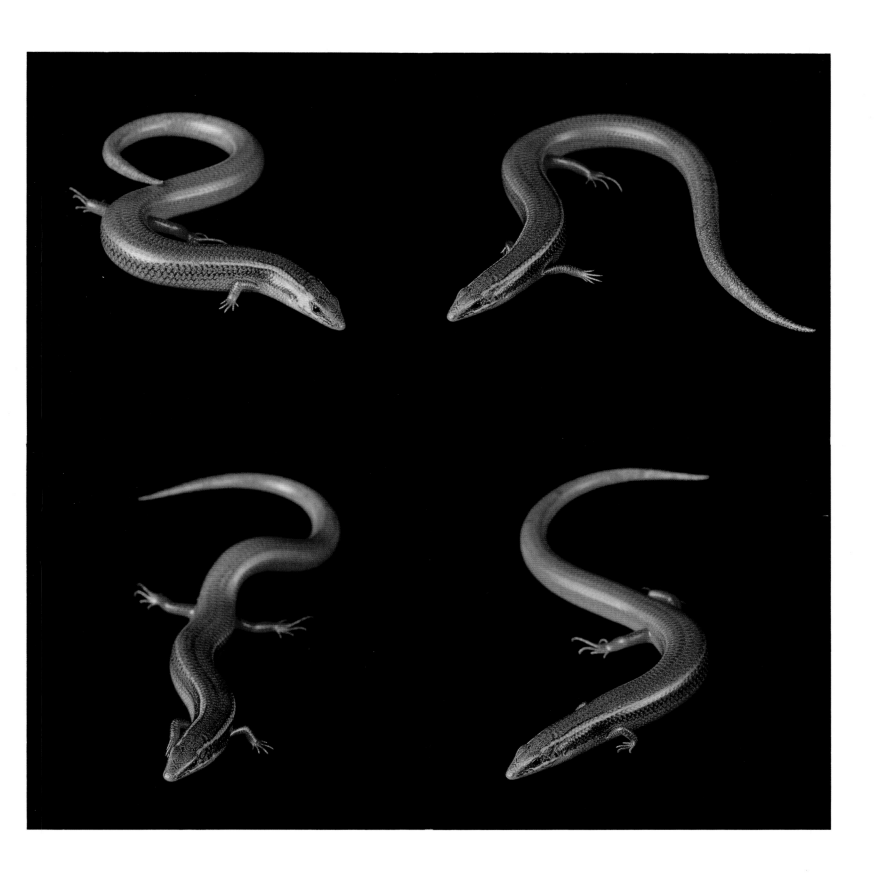

PHOTOGRAPHED
October 12, 1992
University of South Florida
Tampa, Florida

PHOTOGRAPHED
March 19, 1992
Big Pine Key, Florida

Key Tree Cactus

Cereus robinii

STATUS	DATE LISTED	POPULATION	HABITAT	RANGE	THREATS
Endangered	July 19, 1984	Five known populations totaling fewer than 200 individuals	Cactus and hardwood hammock	Florida Keys	Plant collecting, human encroachment

Eastern Indigo Snake

Drymarchon corais couperi

PHOTOGRAPHED
October 14, 1992
Lowry Park
Zoological Gardens
Tampa, Florida

STATUS	DATE LISTED	POPULATION	HABITAT	RANGE	THREATS
Threatened	January 31, 1978	Indeterminate	Mature pine forests, gopher burrows	Florida, Georgia, Alabama	Residential and agricultural development, extermination, illegal collecting

MacFarlane's Four-O'Clock

Mirabilis macfarlanei

PHOTOGRAPHED
June 3, 1993
Pittsburg Landing
Hells Canyon
near White Bird, Idaho

STATUS	DATE LISTED	POPULATION	HABITAT	RANGE	THREATS
Endangered	October 26, 1979	Eight known populations	Loose soils on steep, open slopes	Snake and Salmon river canyons in Oregon and Idaho, and near the Imnaha River in Oregon	Collecting, low numbers, competition from exotic species, insects, grazing, recreational use of habitat

PHOTOGRAPHED
February 21, 1992
Mississippi State University
Starkville, Mississippi

Red-Cockaded Woodpecker

Picoides borealis

STATUS	Endangered
DATE LISTED	October 13, 1970
POPULATION	7,400
HABITAT	Old-growth pine stands
RANGE	Southern U.S., east of central Texas
THREATS	Destruction of old-growth pine forests, isolation of populations, suppression of natural fires

Gopher Tortoise
Gopherus polyphemus

3 weeks old

PHOTOGRAPHED
October 15, 1992
Lowry Park Zoological Garden
Tampa, Florida

STATUS	DATE LISTED	POPULATION	HABITAT	RANGE	THREATS
Threatened in Alabama, Louisiana, Mississippi, Florida	July 7, 1987	Indeterminate	Sandy soils in mature forests	Gulf states from Louisiana to Florida	Habitat fragmentation, egg predation, road-kills

PHOTOGRAPHED
September 14, 1992
Archbold Biological Station
Lake Placid, Florida

Virginia Big-Eared Bat

Plecotus townsendii virginianus

STATUS	Endangered
DATE LISTED	November 30, 1979
POPULATION	10,000
HABITAT	Caves, usually in limestone formations, and surrounding areas of hardwood forest and fields
RANGE	West Virginia, Kentucky, Virginia, North Carolina
THREATS	Human disturbances in caves, loss of foraging habitat, gypsy moths and gypsy moth control programs

PHOTOGRAPHED
August 18, 1992
Franklin, West Virginia

Price's Potato Bean

Apios priceana

PHOTOGRAPHED
August 2, 1992
Missouri Botanical Garden
St. Louis, Missouri

STATUS	DATE LISTED	POPULATION	HABITAT	RANGE	THREATS
Threatened	January 5, 1990	25 known populations totaling 1,700 individuals	Openings in forests or along streams and rivers, disturbed areas	Alabama, Kentucky, Mississippi, Tennessee	Grazing, heavy logging, competition with other plants

Santa Cruz Long-Toed Salamander

Ambystoma macrodactylum croceum

PHOTOGRAPHED
January 17, 1992
Seaside, California

STATUS	DATE LISTED	POPULATION	HABITAT	RANGE	THREATS
Endangered	March 11, 1961	Nine known populations	Shallow, vegetated ponds and surrounding upland chaparral and woodlands	Santa Cruz and Monterey counties, California	Alteration and destruction of breeding ponds and terrestrial habitats, road-kills

Ocelot

Felis pardalis

PHOTOGRAPHED
August 7, 1991
Isis Oasis
Geyserville, California

STATUS	DATE LISTED	POPULATION	HABITAT	RANGE	THREATS
Endangered	July 21, 1982	80–120 in the U.S.	Southwestern brushland	Southern Texas and both coasts of Mexico into Central America	Poaching, agricultural and urban expansion, habitat fragmentation

Wood Stork

Mycteria americana

PHOTOGRAPHED
October 5, 1992
Lowry Park Zoological Gardens
Tampa, Florida

STATUS	Endangered
DATE LISTED	February 28, 1984
POPULATION	5,000 pairs
HABITAT	Subtropical wetlands
RANGE	Southeastern seaboard of U.S.
THREATS	Lack of food from shrinking wetlands

Green Pitcher Plant

Sarracenia oreophilia sarraceniaceae

STATUS	DATE LISTED	POPULATION	HABITAT	RANGE	THREATS
Endangered	September 21, 1979	26 known sites	Acidic soils in woodlands and bogs	Georgia, northern Alabama	Lowered water table, herbicides, suppression of natural fires, agricultural expansion, real estate development, collecting

Texas Blind Salamander

Typhlomolge rathbuni

STATUS	Endangered
DATE LISTED	March 11, 1967
POPULATION	Indeterminate
HABITAT	Underground aquifers
RANGE	Edwards Plateau in Hays County, Texas
THREATS	Groundwater pumping and pollution

PHOTOGRAPHED
November 19, 1992
Cincinnati Zoo
and Botanical Garden
Cincinnati, Ohio

129

PHOTOGRAPHED
October 9, 1992
Lowry Park
Zoological Gardens
Tampa, Florida

West Indian Manatee

Trichechus manatus latirostris

PHOTOGRAPHED
October 27, 1992
Lowry Park
Zoological Garden
Tampa, Florida

STATUS	DATE LISTED	POPULATION	HABITAT	RANGE	THREATS
Endangered	June 2, 1970	1,850	Inlets, river mouths, and ocean near coast	Coastal southeastern U.S.	Collisions with boats, habitat destruction

Welsh's Milkweed

Asclepias welshii

PHOTOGRAPHED
July 8, 1993
Coral Pink Sand Dunes
State Park, Utah

STATUS	DATE LISTED	POPULATION	HABITAT	RANGE	THREATS
Threatened	October 28, 1987	Three known sites totaling 11,000 individuals	Sand dunes	South-central Utah	Off-road vehicles, state priorities

Houston Toad

Bufo houstonensis

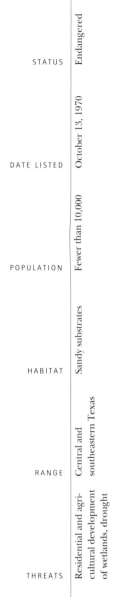

STATUS	Endangered
DATE LISTED	October 13, 1970
POPULATION	Fewer than 10,000
HABITAT	Sandy substrates
RANGE	Central and southeastern Texas
THREATS	Residential and agricultural development of wetlands, drought

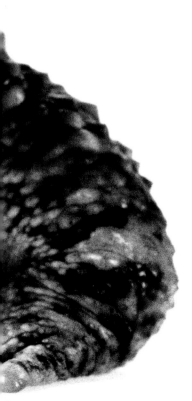

PHOTOGRAPHED
May 2, 1993
Houston Zoo
Houston, Texas

135

3 days old

7 days old

Northern Aplomado Falcon

Falco femoralis septentrionalis

14 days old

17 days old

PHOTOGRAPHED
June 15, 1993
World Center for Birds of Prey
Peregrine Fund
Boise, Idaho

PHOTOGRAPHED
June 14, 1993
World Center for Birds of Prey
Peregrine Fund
Boise, Idaho

STATUS	DATE LISTED	POPULATION	HABITAT	RANGE	THREATS
Endangered	February 25, 1986	Fewer than 70 in U.S.	Desert grasslands and savanna	Texas, New Mexico, southeastern Arizona, primarily Gulf Coast of Mexico	Agricultural development, pesticides

Tennessee Purple Coneflower

Echinacea tennesseensis

STATUS	DATE LISTED	POPULATION	HABITAT	RANGE	THREATS
Endangered	June 6, 1979	Five known populations	Cedar glades in forest openings	Central Tennessee	Residential and industrial expansion, road construction

PHOTOGRAPHED
July 30, 1992
Missouri Botanical Garden
St. Louis, Missouri

141

Northeastern Beach Tiger Beetle

Cicindela dorsalis dorsalis

PHOTOGRAPHED
July 25, 1992
Flag's Pond Nature Park
Calvert County
Maryland

STATUS	DATE LISTED	POPULATION	HABITAT	RANGE	THREATS
Threatened	August 7, 1990	45 known populations	Fine sand beaches	Massachusetts, Maryland, Virginia	Recreational use of beaches, riprap, collecting

143

Sensitive Joint Vetch

Aeschynomene virginica

PHOTOGRAPHED
September 2, 1992
Manokin River Park
Princess Anne
Maryland

STATUS	DATE LISTED	POPULATION	HABITAT	RANGE	THREATS
Threatened	May 20, 1992	10 known populations totaling 7,000 individuals	Freshwater tidal marshes close to estuaries	Middle Atlantic coastal plain	Dredging, water projects, highway construction

Walker's Manioc

Manihot walkerae

PHOTOGRAPHED
April 29, 1993
San Antonio Botanical Garden
San Antonio, Texas

STATUS	DATE LISTED	POPULATION	HABITAT	RANGE	THREATS
Endangered	October 2, 1991	Three known sites totaling 100 individuals (two wild plants in U.S.)	Dry, open brushland	Lower Rio Grande Valley, northeastern Mexico	Agricultural development

PHOTOGRAPHED
June 17, 1992
Warner
New Hampshire

Small Whorled Pogonia

Isotria medeoloides

STATUS	DATE LISTED	POPULATION	HABITAT	RANGE	THREATS
Endangered	October 12, 1982	86 populations totaling 2,600 individuals	Second-growth deciduous or deciduous-coniferous forests	Southern Maine and New Hampshire through the Atlantic seaboard states to northern Georgia and southeastern Tennessee; small colonies in western Pennsylvania, Ohio, Michigan, Illinois, and Ontario, Canada	Collecting, real estate development

Florida Key Deer

Odocoileus virginianus clavium

STATUS	Endangered
DATE LISTED	March 11, 1967
POPULATION	200–400
HABITAT	Pineland and hardwood hammock on subtropical keys and islands
RANGE	Florida Keys, primarily Big Pine and No Name
THREATS	Human encroachment, road-kills, attacks by dogs, drowning of fawns in man-made ditches

PHOTOGRAPHED
March 7, 1992
National Key Deer Refuge
Big Pine Key, Florida

PHOTOGRAPHED
March 4, 1992
Doris Mager Residence
Apopka, Florida

Audubon's Crested Caracara

Polyborus plancus audubonii

STATUS	Threatened in Florida
DATE LISTED	July 6, 1987
POPULATION	500 in Florida
HABITAT	Prairies, brushlands, coastal grasslands
RANGE	Southern Arizona, Texas, Florida, Baja California, mainland Mexico south to Panama
THREATS	Agricultural and residential expansion

STATUS | Endangered

DATE LISTED | June 12, 1989

POPULATION | Indeterminate

HABITAT | Mature forests with deep soils

RANGE | Arkansas, Oklahoma, Rhode Island, Massachusetts

THREATS | Habitat fragmentation

American Burying Beetle

Nicrophorus americanus

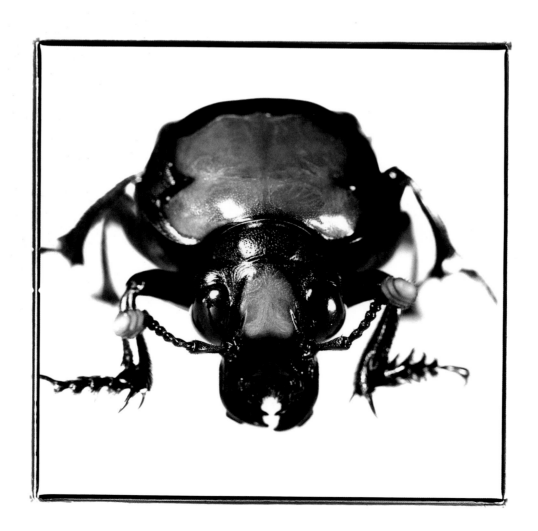

PHOTOGRAPHED
May 20, 1992
Boston University
Boston, Massachusetts

PHOTOGRAPHED
October 29, 1992
University of Florida
Gainesville, Florida

Stock Island Snail

Orthalicus reses reses

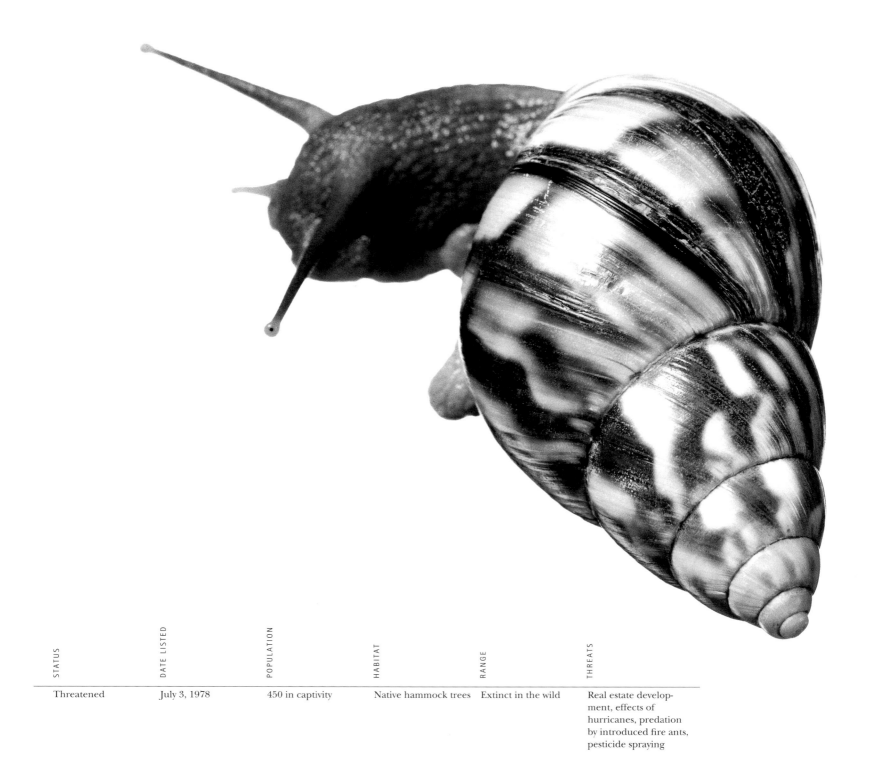

STATUS	DATE LISTED	POPULATION	HABITAT	RANGE	THREATS
Threatened	July 3, 1978	450 in captivity	Native hammock trees	Extinct in the wild	Real estate development, effects of hurricanes, predation by introduced fire ants, pesticide spraying

Scrub Mint

Dicerandra frutescens

STATUS	Endangered
DATE LISTED	November 1, 1985
POPULATION	Four known sites
HABITAT	Well-drained, fine, sandy soils
RANGE	Lake Wales Ridge, central Florida
THREATS	Agricultural and residential development

PHOTOGRAPHED
September 16, 1992
Archbold Biological Station
Lake Placid, Florida

Tooth Cave Spider

Neoleptoneta myopica

PHOTOGRAPHED
May 17, 1993
Tooth Cave
Jollyville Plateau
Texas

STATUS	DATE LISTED	POPULATION	HABITAT	RANGE	THREATS
Endangered	September 16, 1988	Indeterminate	Limestone caves	Jollyville Plateau, Texas	Real estate development, predation by introduced fire ants

Madison Cave Isopod

Antrolana lira

PHOTOGRAPHED
June 20, 1993
Madison Saltpetre Cave
Augusta County, Virginia

STATUS	DATE LISTED	POPULATION	HABITAT	RANGE	THREATS
Threatened	October 4, 1982	Indeterminate	Underground freshwater pools	Augusta County, Virginia	Limited range, pollution

Ash Meadows Sunray

Enceliopsis nudicaulis var. *corrugata*

STATUS	Threatened
DATE LISTED	May 20, 1985
POPULATION	Indeterminate
HABITAT	Dry washes with saline soil, limestone outcrops
RANGE	Ash Meadows, Nevada
THREATS	Off-road vehicles

PHOTOGRAPHED
April 9, 1993
Ash Meadows National
Wildlife Refuge
Nye County, Nevada

Stephens' Kangaroo Rat

Dipodomys stephensi

STATUS	Endangered
DATE LISTED	September 30, 1988
POPULATION	Indeterminate
HABITAT	Native grassland, coastal scrub
RANGE	Perris and San Jacinto valleys and San Luis Rey and Temecula valleys in southern California
THREATS	Agricultural expansion, urban development

PHOTOGRAPHED
March 30, 1992
San Jacinto Wildlife Area
Riverside County, California

Greenback Cutthroat Trout

Oncorhynchus clarki stomias

STATUS	Threatened
DATE LISTED	April 18, 1978
POPULATION	Indeterminate
HABITAT	Mountain streams
RANGE	North-central Colorado
THREATS	Water diversion, mining, competition from non-native trout

PHOTOGRAPHED
November 20, 1991
Saratoga National Fish Hatchery
Saratoga, Wyoming

PHOTOGRAPHED
June 23, 1993
Kansas Ecological Reserves
Rockefeller Native Prairie
Unit #3001
Lawrence, Kansas

Platanthera praeclara

Western Prairie
Fringed Orchid

STATUS	DATE LISTED	POPULATION	HABITAT	RANGE	THREATS
Threatened	September 28, 1989	108 known sites	Remnant native prairies and meadows	Iowa, Kansas, Minnesota, Missouri, Nebraska, North Dakota; Manitoba, Canada	Agricultural development, nectar thievery

Platanthera leucophaea

Eastern Prairie Fringed Orchid

PHOTOGRAPHED
July 3, 1993
Wightman's Grove, Ohio

STATUS	DATE LISTED	POPULATION	HABITAT	RANGE	THREATS
Threatened	September 28, 1989	55 known sites	Remnant native prairies and wetlands	Illinois, Iowa, Maine, Michigan, Ohio, Virginia, Wisconsin; Ontario and New Brunswick, Canada	Agricultural development, suppression of natural fires

Devil's Hole Pupfish
Cyprinodon diabolis

PHOTOGRAPHED
April 13, 1993
Devil's Hole
Ash Meadows National Wildlife Refuge
Nye County, Nevada

STATUS	Endangered
DATE LISTED	March 11, 1967
POPULATION	200–500
HABITAT	Limestone, spring-fed cavern pool
RANGE	Devil's Hole, a single pool in southern Nevada
THREATS	Groundwater depletion

Cheat Mountain Salamander

Pletheodon nettingi

STATUS	Threatened
DATE LISTED	September 22, 1989
POPULATION	72 known populations, most under 10 individuals
HABITAT	Moist, mixed forest above 2,600 feet
RANGE	In and around Monongahela National Forest, West Virginia
THREATS	Removal of forest canopy through logging, recreational development, and mining

PHOTOGRAPHED
August 18, 1989
Elkins, West Virginia

Northern Spotted Owl

Strix occidentalis caurina

PHOTOGRAPHED
October 30, 1991
Corvallis, Oregon

STATUS	DATE LISTED	POPULATION	HABITAT	RANGE	THREATS
Threatened	June 26, 1990	3,500 pairs	Dense old-growth coniferous forest	Northern California north to southwestern British Columbia, Canada	Logging of old-growth forest

Grizzly Bear

Ursus arctos horribilis

PHOTOGRAPHED
September 11, 1991
Olympic Game Farm
Sequim, Washington

STATUS	DATE LISTED	POPULATION	HABITAT	RANGE	THREATS
Threatened	July 28, 1975	Fewer than 1,000 in the U.S.	Wilderness forests, moist meadows, grassland	Wyoming, Montana, Idaho, Washington, Alaska, western Canada	Human encroachment, poaching

Heller's Blazing Star
Liatris helleri

STATUS	DATE LISTED	POPULATION	HABITAT	RANGE	THREATS
Threatened	November 19, 1987	Seven known populations, most with fewer than 50 plants	Rocky outcrops in high-elevation mixed conifer and hardwood forests	Blue Ridge Mountains, North Carolina	Trampling by hikers, natural plant succession, possibly atmospheric pollution

Attwater's Prairie Chicken

Tympanuchus cupido attwateri

STATUS	Endangered
DATE LISTED	March 11, 1967
POPULATION	Fewer than 500
HABITAT	Coastal prairie
RANGE	Narrow stretch of Texas coastal prairie
THREATS	Agricultural and residential development, loss of native grasses to exotic brush

PHOTOGRAPHED
May 23, 1993
Fossil Rim Wildlife Center
Glen Rose, Texas

PHOTOGRAPHED
July 15, 1993
Elk Island National Park
Edmonton, Alberta
Canada

Wood Bison

Bison bison athabascae

STATUS	DATE LISTED	POPULATION	HABITAT	RANGE	THREATS
Endangered in U.S., threatened in Canada.	June 2, 1970	2,985	Boreal forest and prairies	Yukon, Northwest Territories, Alberta, Manitoba, Canada	Hybridization with plains bison, low numbers, disease

Bald Eagle

Haliaeetus leucocephalus

STATUS	DATE LISTED	POPULATION	HABITAT	RANGE	THREATS
Threatened or endangered in lower 48 states, unclassified in Alaska	March 11, 1967	Fewer than 5,000 in lower 48 states	Lakes, rivers, and other wetlands	Alaska, lower 48 states, Canada	Poaching, toxic contamination, wetland conversion

PHOTOGRAPHED
August 16, 1990
San Francisco Zoological Gardens
San Francisco, California

Hibiscadelphus distans
Kauai Hau Kuahiwi

PHOTOGRAPHED
February 10, 1993
National Tropical Botanical Garden
Kauai, Hawaii

STATUS	DATE LISTED	POPULATION	HABITAT	RANGE	THREATS
Endangered	April 29, 1993	Two known sites totaling 100 individuals	Open, dry forest	Waimea Canyon, Hawaiian Island of Kauai	Feral goats, competition from introduced vegetation, trampling by hikers, predation by insects

PHOTOGRAPHED
February 17, 1993
Mauna Kea Forest Reserve
Hawaii, Hawaii

STATUS | Endangered

DATE LISTED | March 21, 1986

POPULATION | 1,000

HABITAT | Barren alpine scrub high on volcanic slopes

RANGE | Mauna Kea Forest Reserve, Island of Hawaii

THREATS | Wild sheep and feral goats, predatory insects, limited distribution

Argyroxiphium sandwicense ssp. *sandwicense*

Ahinahina

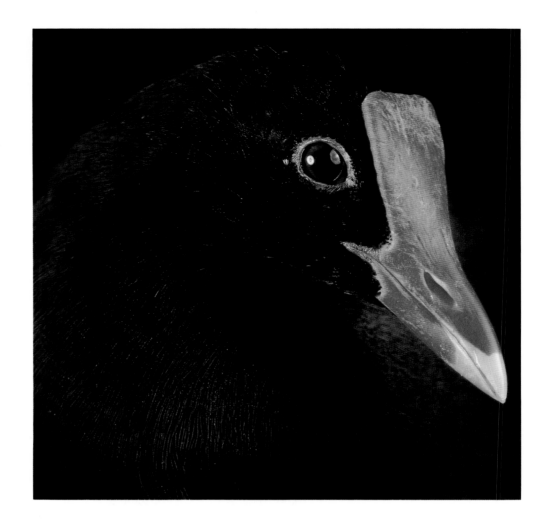

STATUS	Endangered
DATE LISTED	March 11, 1967
POPULATION	750
HABITAT	Freshwater marshes, grassy wetlands
RANGE	Hawaiian Islands of Kauai and Oahu
THREATS	Filling and degradation of wetlands, agricultural and residential development, predation by mongooses and feral cats

ʻAlae ʻUla

Gallinula chloropus sandvicensis

PHOTOGRAPHED
January 23, 1993
Honolulu Zoo
Oahu, Hawaii

PHOTOGRAPHED
October 1, 1991
Sea World of California
San Diego, California

Nesochen sandvicensis
Nene

STATUS	DATE LISTED	POPULATION	HABITAT	RANGE	THREATS
Endangered	March 11, 1967	500	Rugged, sparsely vegetated lava flows	Hawaiian Islands of Hawaii and Maui	Low reproduction, predation, livestock overgrazing

PHOTOGRAPHED
January 26, 1993
Honolulu Zoo
Oahu, Hawaii

Koloa Maoli

Anas wyvilliana

STATUS	Endangered
DATE LISTED	May 11, 1967
POPULATION	2,500
HABITAT	Mountain streams, montane bogs and ponds, other wetlands from sea level to over 3,500 feet
RANGE	Hawaiian Islands of Kauai, Oahu, and Hawaii
THREATS	Development of wetlands, hybridization with non-native species, introduced predators

PHOTOGRAPHED
January 26, 1993
Waikiki Aquarium
Honolulu, Hawaii
February 3, 1993
Sea Life Park
Waimanalo, Oahu
Hawaii (pups)

STATUS	DATE LISTED	POPULATION	HABITAT	RANGE	THREATS
Endangered	November 23, 1976	1,500	Submerged reefs and sheltered beaches on islands and atolls	Northwestern Hawaiian Islands from Nihoa to Kure	Disease and injury, commercial fishing: competition for prey, entanglement in nets

Lasiurus cinereus semotus
Hawaiian Hoary Bat

STATUS	DATE LISTED	POPULATION	HABITAT	RANGE	THREATS
Endangered	October 13, 1970	Indeterminate	Roosts in trees or rock crevices, forages in forest clearings, open fields, and occasionally over open ocean	Hawaiian Islands, chiefly Hawaii and Kauai, occasionally Oahu and Maui	Residential and urban development, conversion of rain forest into plantations

Laysan Duck

Anas laysanensis

STATUS	DATE LISTED	POPULATION	HABITAT	RANGE	THREATS
Endangered	March 11, 1967	500	Dense vegetation near water	Hawaiian Island of Laysan	Low numbers

PHOTOGRAPHED
January 24, 1993
Honolulu Zoo
Oahu, Hawaii

201

ʻIhiʻIhi

Marsilea villosa

STATUS	Endangered
DATE LISTED	June 22, 1992
POPULATION	Three sites totaling fewer than 1,000 individuals
HABITAT	Vernal pools in dry lowland areas
RANGE	Hawaiian Islands of Oahu and Molokai
THREATS	Drainage of wetlands

PHOTOGRAPHED
January 27, 1993
Waimea Arboretum
and Botanical Garden
Oahu, Hawaii

Na'u

Gardenia brighamii

STATUS	Endangered
DATE LISTED	August 21, 1985
POPULATION	15
HABITAT	Dry lowland forests
RANGE	Hawaiian Islands of Lanai, Oahu, and Molokai
THREATS	Grazing, feral animals, rats

ʻAeʻo

Himantopus mexicanus knudseni

STATUS	Endangered
DATE LISTED	October 13, 1970
POPULATION	1,000–1,500
HABITAT	Coastal wetlands
RANGE	Hawaiian Islands, chiefly Maui and Oahu
THREATS	Filling and degradation of wetlands, changes in agricultural crops, predation by mongooses and feral dogs and cats

PHOTOGRAPHED
January 24, 1993
Honolulu Zoo
Oahu, Hawaii

Achatinella fuscobasis

Achatinella decipiens

Oahu Tree Snails
Achatinella spp.

STATUS	DATE LISTED	POPULATION	HABITAT	RANGE	THREATS
Endangered	August 31, 1981	19 species, each with 20–1,000 individuals	Tropical upland forest	Waianae and Koolau ranges on the Hawaiian Island of Oahu	Predation by non-native carnivorous snails, rats, and flatworms, alien tree species replacing native plants

PHOTOGRAPHED
January 31, 1993
University of Hawaii
Honolulu, Hawaii

Achatinella mustelina

PHOTOGRAPHED
February 12, 1993
Panaewa Rainforest Zoo
Hawaii, Hawaii

ʻIo

Buteo solitarius

STATUS	Endangered
DATE LISTED	March 11, 1967
POPULATION	2,700
HABITAT	Open or parkland forests and dense rain forests from near sea level to 8,500 feet
RANGE	Island of Hawaii
THREATS	Nest disturbance, agricultural and urban expansion

Lanai

STATUS	DATE LISTED	POPULATION	HABITAT	RANGE	THREATS
Endangered	September 26, 1986	10 known populations totaling 400 individuals	Eroded slopes, gullies, lava fields	Hawaiian Islands of Maui, Lanai, and Hawaii	Grazing by cattle, goats, and deer, real estate development

PHOTOGRAPHED
January 28, 1993
Waimea Arboretum
and Botanical Garden
Haleiwa, Hawaii

Maui

Lanai

Abutilon menziesii
Ko'oloa'ula

Maui

PHOTOGRAPHED
February 22, 1993
Hawaii Department of Land
and Natural Resources
Forestry Nursery
Maui, Hawaii

PHOTOGRAPHED
February 23, 1993
Hawaii Department of Land
and Natural Resources
Maui, Hawaii

Cooke's Kokio

Kokia cookei

Kokia cookei

GRAFT - - - - - - - ▸

Kokia drynarioides

STATUS	DATE LISTED	POPULATION	HABITAT	RANGE	THREATS
Endangered	October 30, 1978	Fewer than 50 clones	Dry and semidry lowland forest	Hawaiian Island of Molokai	Low numbers, lack of rootstock

Species Profiles

Ae'o

*Himantopus
mexicanus knudseni*

The elegant ae'o, or Hawaiian stilt, stands balanced on one of its foot-long, pink legs, dozing in the thick of a wildlife refuge. With the dark coloration along its back—black in the male, dark brown in the female—and smooth white breast feathers, the stilt has an air of refinement, as if it were decked out in formal attire for its flights among the Hawaiian Islands.

As the ae'o stalks the coastal wetlands looking for food, its narrow, pointed beak is perfect for snatching up shore flies, water boatmen, and other aquatic insects, small fish, and crabs. At sunset, accompanied by loud cries, the stilts take flight en masse and return to the small islands within the wetlands where they nest.

The nests are often little more than scrapes in the ground lined with twigs or other debris. They usually contain four smoky gray eggs, speckled with chocolate brown. The eggs are vulnerable to mongooses, and the Hawaiian stilt tries to distract them, and other predators, by feigning injury and sounding noisy yelps. Stilts incubate their eggs for about 24 days until the young emerge, coated in brown and black down and ready to forage for food as soon as their feathers have dried. Within about 50 days, the young are able to fly.

Volcanic islands are not known for their wetlands. Those on the Hawaiian Islands were supplanted by taro patches and later by rice fields. Although the building of high-rise hotels and golf courses threatens the stilt's habitat, the species is still present on all the islands of its historic range. Since the stilt was designated as endangered, its population has risen from the alarmingly low count of 253 in 1960 to between 1,000 and 1,500 birds. Maui and Oahu host approximately 65 percent of the population, but stilts also live on the islands of Molokai, Hawaii, Kauai, and Niihau.

The U.S. Fish and Wildlife Service has supervised vigorous conservation measures, including setting aside valuable wetland areas. Kanaha and Kealia ponds on Maui were the first state sanctuaries, and five national wildlife refuges, including the 917-acre Hanalei National Wildlife Refuge on Kauai, have been established. A Hawaiian Waterbirds Recovery Team is attempting to increase the number of stilts to 2,000 by 1996. It has also developed artificial nesting islets and floating nest structures to bolster the number of these beautiful birds.

Ahinahina

*Argyroxiphium
sandwicense* ssp. *sandwicense*

"Truly superb, and almost worth the journey of coming here to see it on purpose." So Scottish botanist James Macrae described the ahinahina, or Mauna Kea silversword, when he came across it in 1825. At that time, above the tree line, over 9,000 feet up the volcanic slopes of the Big Island, this silversword was abundant. Even as recently as 50 years ago, a traveler complained that he was temporarily blinded by the sunlight reflected off these broad-leaved plants as he walked through the upper Wailuku River basin.

Unlike most Hawaiian floras which appear to have their roots in Asia, the Hawaiian silverswords, five closely related plants, have recently been found to be descended from the North American tar-weeds. The ahinahina (gray-gray) grows in the form of one or more rosettes about two feet in diameter, made up of one-foot-long leaves. After five to 15 years, the plant produces a flowering stalk that grows rapidly up to eight feet in height and which bears pinkish flowers. The flowers then turn to seeds, and plants with only a single rosette die.

Incapable of self-fertilization, the ahinahina relies for reproduction on the cross-pollination between two plants. At one time the number of silverswords had dropped below 100, and with so few plants extant, the likelihood of cross-pollination occurring was small indeed, especially as the insect pollinators, primarily moths, are also on the decline. With the silversword's status already very precarious, mouflon sheep from the Mediterranean were introduced in the late 1950s for hunting. Hunters like mouflon sheep because they do not flock. Mouflon sheep liked Mauna Kea silverswords—what few were left.

Methodical propagation of silverswords in the Mauna Kea Forest Reserve over the last 20 years is bearing fruit, and a second generation of cultivated plants has been transplanted into the wild. In the meantime the sheep, as well as feral goats, have been removed from the reserve, and the Mauna Kea silversword's future is looking brighter.

'Alae'Ula

*Gallinula chloropus
sandvicensis*

Native Hawaiians honor this secretive bird, for in their lore the 'alae'ula stole a blazing brand from the gods and gave fire to the people. One word of its Hawaiian name, *'alae*, signifies the red shield above the bird's bill, a lasting mark from that fiery flight.

Also called the Hawaiian common moorhen, it is a subspecies of the common moorhen found in North America. This foothigh water bird with garish green legs probably evolved from stray island colonists and is nonmigratory. It was once common on all the main Hawaiian Islands except Lanai and Kahoolawe. By the middle of this century, the 'alae'ula had disappeared from Hawaii and Maui, and estimates placed the total population at only 60 birds.

Today, populations on Kauai and Oahu, the last islands where it is found, seem stable, though the bird's reliance on aquaculture fields and other remnant wetlands means that its survival may hang on the protection of such freshwater habitat, where the 'alae'ula feeds on algae, seeds, aquatic insects, and mollusks.

The 'alae'ula nests in dense vegetation on the edges of marshes. When water levels are high and plants sport luxuriant growth, the bird weaves reeds together into a platform nest. Breeding occurs mainly from March through August, and clutch size varies from five to 13 eggs. The ground-level nest attracts introduced predators like mongooses and feral cats, and exotic plants such as water hyacinth and mangrove choke the native plant species used in nest building.

Since 1972, the 'alae'ula, along with the endangered koloa maoli (page 228), ae'o (above), and other species, have been protected at Hawaii's first water-bird refuge—Hanalei National Wildlife Refuge on Kauai, 917 acres that escaped draining and filling. On Oahu, the James Campbell and Pearl Harbor national wildlife refuges retain critical habitat for these birds. Due to the 'alae'ula's reclusive nature, the current estimate of 500 on Kauai and 250 on Oahu is likely to be low.

Aleutian Canada Goose

Branta canadensis leucopareia

Thousands of Aleutian Canada geese once bred in Alaska throughout the central and western Aleutian islands, west to the Commander Islands and southwest to the Kuriles. These geese are smaller and somewhat darker than most other subspecies of Canada goose, and all adults have a distinctive white neck ring. The Aleutian Canada goose carved out its own particular niche on those small, rugged islands that border the north Pacific Ocean. They evolved on islands free of mammalian predators, where their main enemies were gulls, eagles, and jaegers. When fur traders started bringing arctic foxes and red foxes to the islands in the 1750s, the young geese were easy prey. So, too, were the adults, which become flightless for a few weeks each year during their late summer molt.

A few years after foxes were introduced to each island, most of that island's birds, and all of its Aleutian Canada geese, were wiped out. The authorities regulating the fur traders may not have realized how destructive the foxes could be, for as late as the 1930s, foxes were still being introduced to previously fox-free islands. For instance, foxes were first released in 1923 on Agattu, where thousands of Aleutian Canada geese were known to nest. Fifteen years later there were none. Those geese that were able to circumvent the foxes and make their annual southern migration through Canada often met with sport hunters in California and Oregon. By the late 1930s the Aleutian Canada goose, which had once flown overhead in vast numbers, their calls loudly signaling the coming of spring and winter, had nearly vanished.

For nearly 30 years, there was no confirmed sighting of the goose. Then, in 1962, the bird was rediscovered on remote Buldir Island, which had escaped the intrusion of foxes. Biologists were able to trap and transplant geese to four other Aleutian Islands, now cleared of foxes. In 1975 hunting of the goose was banned, and by 1992 the bird's population, which once stood at 790, had climbed to over 8,000. Its status was changed from endangered to threatened.

American Burying Beetle

Nicrophorus americanus

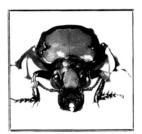

As with other members of the genus *Nicrophorus* (meaning "bearer of the dead"), the distinctive characteristic of the American burying beetle is that its life cycle centers on two animal carcasses: one in which it is born and one in which it rears its young. Sensitive chemical receptors on the beetle's antennae locate a carcass, generally a bird or a rodent such as the cotton rat, weighing three to six ounces—substantially more than the beetle. If the beetle is a female, she generally sets about burying the carrion immediately. A male may first look for a female for help. If, as often happens, more than one male or female shows up, drawn to the smell of the carcass, they must fight for possession, males with males, females with females, until only one of each, usually the largest, remains.

If the body is not already in a suitable burying place, the pair lie on their backs under the animal and push it along with their feet, sometimes a foot or more from where they found it. Once they reach soft ground, they bury the carcass two to four inches below the surface, hidden from rivals and other scavengers, and squeeze it into a ball. Working together, they hollow out a brood cell in the carcass, then mate.

The eggs are laid in a tunnel adjacent to the carrion, and both parents stay until the larvae hatch and start eating the carcass. With food so readily available in such abundance, the larvae increase their weight 10-fold each day. After a few days the male usually leaves, but that he stays at all for the hatching and critical first days of his offspring is unusual among nonsocial insects. Studies have shown that if the male leaves or is removed from the nest too early, his position is likely to be usurped by another male, who will kill the existing larvae and mate with the female.

That, until recently, the only two known sites of the American burying beetle were on Block Island, off Rhode Island, and in Oklahoma, is an indication of how widespread the species once was. Why the decline of the American burying beetle has been so much more precipitous than that of related species is unknown. It may be partly because the relatively large animals on which the American burying beetle depends are themselves victims of habitat loss. With fewer than 1,000 beetles left in the Northeast, scientists have begun breeding them in laboratories with a view to establishing new populations. Of 25 pairs introduced onto Penikese Island off the Massachusetts coast in 1990, at least a dozen were found to have healthy offspring.

For a creature with a life expectancy of one year, dependent upon the capricious chance of finding recently deceased carcasses, a thousand instances are precious few. Yet what a loss to the world's intricate diversity were the American burying beetle to leave us.

American Crocodile

Crocodylus acutus

American crocodiles, despite their reputation, are not aggressive toward humans. That characteristic belongs to the more numerous American alligators, which live in adjacent, less saline habitats and with which crocodiles are often confused. The jaws of crocodiles are more tapered than those of alligators, giving them a more triangular-shaped snout. Moreover, the fourth tooth in the lower jaw of the crocodile is more distinct: it is large and always visible, even when the mouth is closed.

Crocodiles are rarely seen even when they live close to humans. They rest throughout the day beside mangrove-lined estuaries, preferring brackish, slow-moving, warm water with a fairly narrow temperature range. At night they move out into the creeks and canals and into coastal waters to feed on fish, crabs, birds, and turtles.

The female is choosy about where she digs her nest, for her eggs face the dual threats of drowning and dessication. Suitable places are rare, and a crocodile may come back to the same spot each year, and may even, unwittingly, share a nest with another female. She generally lays up to 40 eggs, close, but not too close, to the water's edge, then covers them with loose sand. The parents never move far from the nest, and the mother defends it against intruders. Nevertheless, predators, notably raccoons, sometimes manage to dig up and make a meal of the embryos. If the eggs survive, 60 to 75 days later, depending upon how warm the spring, the mother will hear the juvenile chirping. It is time to dig up the eggs, break open the still-hard shells, and carry the hatchlings to the water.

The hatchlings are still at risk, especially from other crocodiles or from alligators. Only after they are one or two years old and over two feet long—about the size of the crocodile in the photograph—are they fairly free of predation. Then, if they can survive murderous territorial battles with other crocodiles, and do not get run over while taking advantage of the retained heat of a roadway to keep warm, they may live to a ripe old age. Crocodiles over 50 years old are known, and some may live much longer.

The prognosis for the American crocodile is not good. Other populations of this species in the West Indies and Central America are even more threatened by human expansion than those in Florida. Their survival will require eternal vigilance.

Arizona Agave

Agave arizonica

Every 10 to 30 years the Arizona agave stretches forth its flower stalk. The stalk, or inflorescence, grows rapidly, and within a few weeks it reaches 12 to 16 feet into the hot desert air. Starting from the base, one by one the small, tubular, pale yellow flowers open. Each flower lasts for three to five days before it fades. For a month while the flowering progresses, bees and other insects hover about the agave, diving into the nectar-laden flowers and emerging covered with pollen.

The probability that a bee will ever transfer that pollen to another Arizona agave is virtually nil. There are only 60 known clones in the wild, and though they are found only in a few mountain ranges of Arizona, they are widely separated. Pollination would also require the unlikely event of two nearby plants flowering at the same time. The prolonged period between flowerings of the few remaining Arizona agave has even made it difficult for scientists to conduct reproductive biology experiments, and some suspect that the plant may be an infertile hybrid.

The Arizona agave does reproduce asexually by underground roots, or rhizomes, which form offsets around the mother plant. Botanists refer to this cluster of identical plants as a "mother and pups," or a clone.

For most of its life, when the Arizona agave is not flowering, it is a small, rather unassuming plant, reaching, at maturity, about a foot high and a foot in diameter. The green coloration of the thick, dark green leaves gives way to a mahogany red margin along their often toothed edges.

The insignificant appearance of the Arizona agave, together with its scarcity and the inaccessibility of its habitat, has largely kept it safe from collectors. But the agave does not escape the attention of browsing cattle and deer. They forage on the sugar-filled inflorescence as it emerges, before the stalk has a chance to flower and produce seed. Many of the clones are fenced off as scientists work to increase their numbers.

Ash Meadows Sunray

Enceliopsis nudicaulis
var. *corrugata*

Only 75 miles northwest of Las Vegas, Ash Meadows harbors more unique plants and animals—26 including the Devil's Hole pupfish (page 220)—than any other place in the continental United States. Consequently, it has been called a mainland Galápagos. A series of deepwater springs on 50,000 acres of desert uplands straddling the California-Nevada border formed millions of years ago. A cool, moist oasis surrounded by the Mojave Desert, Ash Meadows nourished an ecosystem that evolved in isolation as extreme as that on a midocean island.

The sunray is one of seven unique Ash Meadows plants on the endangered or threatened lists (all but one of the listed plants for all of Nevada). A member of the sunflower family, the sunray is the showiest of the Ash Meadows plants. Relatively common in the uplands where it occurs, this perennial with strongly ruffled leaves grows in clumps four to 16 inches wide and two feet tall, with several flowering stalks topped by two-inch flowers. In spring, the aptly named flowers burst yellow from the bare ground.

The springs lured Shoshone Indians who camped at Ash Meadows, and later drew white settlers who grew crops. Upon attaining statehood in 1864, Nevada gave land rights along the spring channels to private owners. That act began a long period in the region's history when the springs were ditched and diverted to irrigate alfalfa and provide pasture for cattle.

Then in 1984, The Nature Conservancy bought 12,654 acres at Ash Meadows, on the site of an aborted residential resort, and transferred them to the U.S. Fish and Wildlife Service. The entire Ash Meadows National Wildlife Refuge encompasses about 21,000 acres, including Bureau of Land Management property, and protects all of the sunray's critical habitat in the southern and eastern portions of Ash Meadows. Some sunray habitat is on private land.

The biggest threat to the sunray comes from illegal off-road vehicle (ORV) use on the refuge. In recent years, miles of gravel roads were built, crisscrossing critical habitat for the plant and providing access to the ORVs. The habitat cannot be fenced off and is difficult to patrol, so signs discouraging drivers provide the only constant vigil.

Because it occurs in the uplands, the sunray has at least been spared the threat of competition from exotic plants and weeds that plague species in the refuge's wetter regions. For instance, saltcedar, or tamarisk, a Mediterranean native planted as a windbreak in the arid West, invades stream and river corridors and quenches its enormous thirst by tapping deep into the water table.

Attwater's Prairie Chicken

Tympanuchus cupido attwateri

For thousands of years the Attwater's prairie chicken depended on buffalo to graze Texas's coastal tall-grass prairies. The great woolly mammals would pass through each year, grazing on the little blue stems, Indian grass, big blue stems, and switch grass. The prairie chicken requires these grasses, in the right concentrations, for nesting and cover. Too much grass, and the birds can't move around to forage for food. Too little, and they have no place to lay their eggs or hide from predators.

In the spring, love-starved males gather on open spaces, called "booming grounds," to strut and fight over the females in attendance. After a vigorous stamping with their feet, they send out deep booming noises from inflated air sacks on the sides of their necks.

Declining habitat resulted in a reduced chicken population as farmers began to replace much coastal prairie with agricultural fields, mainly rice. Remaining prairie was often overgrazed by cattle, leaving no place for the hens to lay their eggs and no place for the birds to take shelter during the winter months. Where the cattle overgraze, the native grasses also tend to be overrun by exotic plants that are of no use to the birds. Some of these plants are shrubs—these used to be controlled by fires set by lightning or by Native Americans and later by some ranchers. The growth of brush creates perches and resting sites for owls and hawks, enabling these predators to move into prairie chicken territory where they feed on the young. Urban sprawl, especially around Houston, has also replaced much of the prairie

chicken's habitat. Together, these threats have reduced a population that probably exceeded a million in the 1800s to fewer than 500 birds.

Although their numbers continue to decline, a project funded by the U.S. Fish and Wildlife Service has succeeded in breeding Attwater's prairie chickens in captivity. Another captive-breeding program, at Texas A&M University, is funded by the Texas Parks and Wildlife Department (TPWD). The TPWD has also initiated a program to encourage private landowners to restore their prairie chicken habitat by removing aggressive exotic plants and by managing cattle grazing to preserve native grasses at a level conducive to the bird's requirements.

Audubon's Crested Caracara

Polyborus plancus audubonii

From a high perch at dawn or sunset, this tropical raptor throws back its head, burying the neck feathers in its shoulders, and gives a cackling call—a call that inspired Paraguay's Guarani Indians to name the bird *cara-cara*. The caracara, sometimes called the Mexican eagle, appears on the national seal of Mexico, a rattlesnake dangling from its beak. Actually a member of the falcon family, the caracara shares many characteristics with the vulture family. Its featherless face and long legs and neck assist the caracara in eating carrion; the bird places one foot on a carcass and tears off flesh with its powerful beak. An opportunist with an aggressive appetite, the caracara hunts a variety of small prey and even harasses pelicans and gulls until they relinquish their caught fish.

The two-foot-long bird mates for life, and each pair have a large home range. They build a bulky nest of branches lined with sticks, roots, and grass atop trees or on cliffs, where the female lays a clutch of two or three eggs. Young are fed fresh meat rather than regurgitated fare. John James Audubon first saw the species in Florida in 1831 when he watched a caracara feast with vultures on a dead horse near St. Augustine. His assistant shot the bird and Audubon began painting it, watching the colors of the feathers fade as he worked.

Today, the caracara is gone from St. Augustine and fares poorly throughout Florida. Its statewide population has dropped by more than two-thirds since the turn of the century, and its range is still shrinking. The

bird's recovery plan emphasizes purchasing or otherwise protecting parcels of critical habitat. Most caracaras occur on private land around Kissimmee Prairie north of Lake Okeechobee, where they can be seen patrolling highways in search of road-killed animals, often becoming casualties themselves.

Bald Eagle

Haliaeetus leucocephalus

Hurtling through space, a pair of bald eagles sometimes lock talons in midair, as they consummate courtship rituals of high-speed ascents and somersaulting falls. Despite Benjamin Franklin's preference for the wild turkey, it is no wonder that this beautiful bird inspires a longing for majesty and liberty and that it was chosen as the national bird by the U.S. Continental Congress in 1782.

Two hundred years later, the bald eagle inspires a profound nostalgia for the open spaces and vital wetlands of centuries past. It was once plentiful over nearly the entire continent, but beginning in the 19th century, habitat loss and hunting have steadily reduced bald eagle populations. By the 1850s, for instance, the once-great popula-

tions of bald eagles in the Adirondack Mountains had been extinguished by the logging of their lake and riverside pine forest habitat.

In the 1940s, the reckless use of the pesticide DDT was common, and the poison quickly worked its way up the food chain in accumulations of higher and higher doses. As eagles were hit by these potent doses of DDT in the fish they ate, the eggs they laid had shells so thin that many embryos became dehydrated and never hatched. Eagle populations plummeted across the continent, and the bird disappeared altogether from many states.

The 1972 U.S. ban on most uses of DDT marked the beginning of a model recovery

program pairing habitat protection with determined efforts to reintroduce eagles throughout their historic range. By 1981, the nesting population in the lower 48 states had doubled, and small eagle populations had been restored to sites around the country, prompting the U.S. Fish and Wildlife Service to downlist the bald eagle populations in Minnesota, Wisconsin, Michigan, Oregon, and Washington from endangered to threatened status.

The eagle still faces threats, however, as illegal shooting continues and relentless land development edges the bird out of its remaining habitats.

Black-Footed Ferret

Mustela nigripes

The black-footed ferret once ranged throughout the North American plains. By 10 years ago most biologists considered it doomed.

This ferret's survival is inextricably tied to the welfare of prairie dogs. They are its meat and potatoes. They even provide shelter: the ferret, having consumed a family of prairie dogs, often moves into the abandoned burrow. This is an unusual predator-prey relationship, for the predator is depending for its survival on a species that's as big or bigger than it is.

Prairie dogs live in large communities, "dog towns," which they like to build in the grassland country east of the Rocky Mountains. That is good grazing land, and

the cattle ranchers who regard these rodents as competition have reduced their numbers to less than 5 percent of the original population. As goes the prairie dog so goes the black-footed ferret. In 1986, with the population in rapid decline, the last remaining ferrets biologists could find, all 18 of them, were trapped and put into the captive-breeding station in Wheatland, in southeastern Wyoming.

Though ferrets are tough animals and strong fighters, they are susceptible to disease, so the breeding station is kept spotless. No one is allowed into the station who does not first don a mask and surgical gloves to prevent the ferrets from succumbing to colds and flu. Such precautions have worked, and

in 1991, with the population back up close to 200, the first successful reintroduction of the creatures back into the wild was completed.

The black-footed ferret is still the rarest mammal in North America. If it is to survive, it will be in carefully managed areas, and it will depend upon the willingness of ranchers to allow occasional dog towns to flourish adjacent to their grazing lands.

Approxi-mately 10 percent of the reintroduced ferrets survived their first year in the wild, cor-responding to the normal survival rate for ferret populations. The identification and protection of dog towns suitable as reintroduction sites now appear to be the last major obstacles to a successful recovery program.

Blue-Tailed Mole Skink

Eumeces egregius lividus

Protective land management practices sometimes do more harm than good. Suppressing fires, for instance, keeps some creatures alive today, but in the long run may lead to a buildup of combustible materials and eventually to bigger, uncontrollable fires that a habitat is unable to withstand. Many species, and entire ecosystems, rely on periodic fires to keep their various systems in balance. The Florida scrub community is certainly one of these. Before human intervention, it was periodically set ablaze by lightning, purged of excess of undergrowth, and nourished by the recycled nutrients.

The blue-tailed mole skink and the sand skink (*Neoseps reynoldsi*) are two of the many animals and plants in the scrub community whose habitat has been demeaned by fire suppression and also has been reduced by agricultural development. Though these two endangered skinks occupy the same areas of Florida, they are not really competitors since they concentrate their foraging efforts on different altitudes.

The blue-tailed mole skink, which has functional legs, spends much of its time aboveground looking for insects such as spiders, roaches, and crickets, whereas the sand skink, whose legs are so small they are almost useless for walking on the surface, spends most of its time foraging under the sandy soil for smaller fare such as tiny termites, beetle larvae, and ant lions.

The blue-tailed mole skink occurs only in three counties in central Florida, and much of its habitat has been transformed into citrus groves. At least 14 other species of animals and plants endemic to this part of Florida are on the endangered species list, and far more than that are threatened with extinction. In the last 50 years more than 80 percent of the skink's habitat has been altered beyond usefulness to the creature.

Biologists say the blue-tailed mole skink's survival will depend both on the purchase and protection of its remaining habitat and on the appropriate management of those reserves, including periodic controlled burning.

Boulder Darter

Etheostoma wapiti

Like its more celebrated cousin, the snail darter, which threatened to halt construction of Tellico Dam on the Little Tennessee River, the boulder darter has found its nemeses in dams and poor land use. One of the most imperiled fishes in the southeastern U.S., the boulder darter inhabits the Elk River in Tennessee and Alabama and the lower reaches of two tributaries, Indian and Richland creeks. Its range on the Elk has declined substantially, and the darter no longer occurs on the Tennessee River or Shoal Creek. Only eight small, isolated populations are known.

Darters are a diverse and fascinating group of small perches found only in North America. Most common in the eastern U.S., darters number about 140 species, a fifth of all freshwater fish species in North America excluding Mexico. In spring, their mating colors and habits rival those of the gaudiest butterflies and coral reef fishes.

The three-inch-long boulder darter's common name comes from its preferred habitat of boulder substrate. At two sites lacking boulders, the darters use the rubble of a collapsed stone bridge and a spillway dam. The boulder darter spawns in cavities beneath boulders. Males select narrow, horizontal cracks and guard the eggs wedged there by females. Both sexes have multiple mates, but little else has been learned about the boulder darter's life history.

The construction of Tims Ford Dam on the Elk River has threatened boulder darter populations. Fluctuating water levels brought by daily releases below the dam cause dramatic changes in the darter's specific habitat. Other darter populations were probably lost with the building of two dams on the Tennessee River.

Many darters are sensitive to water pollution and sedimentation. On the Elk River, pollution from herbicides and insecticides presents a problem, and sedimentation has increased after farmers removed riparian vegetation along significant stretches of the river. Industrial and agricultural pollution from Florence, Alabama, helped remove the boulder darter from Shoal Creek. But the pollution has declined in recent years, and boulder darters may be restored to Shoal Creek, if enough adults can be collected to start a captive-spawning and -rearing program.

California Condor

Gymnogyps californianus

With a wingspan over nine feet, the California condor is the largest flying bird in North America. In prehistoric times the condor occurred along the southern coast as far east as Florida, and historically it soared from British Columbia south to northern Baja California. By the 1940s, however, the condor's numbers had sharply declined, and the bird survived only in the mountains of southern California's Los Padres National Forest.

In 1985, after six of the last 15 wild birds had disappeared mysteriously during the previous winter, biologists decided to take a desperate step. The last wild California condor was captured and taken into captivity in 1987, when the species numbered only 27 individuals. A controversial captive-breeding and reintroduction program run by the U.S. Fish and Wildlife Service, California Department of Fish and Game, and Los Angeles and San Diego zJ219

oos hatched its first chick the following year. Four, eight, and 12 chicks, respectively, hatched successfully in the subsequent three years. Another 12 hatched in 1992 and 15 in 1993.

Meanwhile, the recovery team released a crew of female Andean condors to test the skies around the Sespe Condor Sanctuary in the Los Padres National Forest. Then, in January 1992, after nearly five years when the world went without wild California condors, the first captive pair were freed from a guarded, cliff-top enclosure 50 miles northwest of Los Angeles. The young female, Xewe, is the offspring of AC-9, the last condor taken into captivity in 1987. Xewe survived being shot at in July 1992, but her male partner, Chocuyens, was found dead nine months after the release, poisoned by ethylene glycol, possibly from antifreeze.

Those incidents underscore the uphill road to recovery that the California condor faces. Condors feed solely on carrion, the remains of dead animals. In the Pleistocene, they scavenged the carcasses of mammoths and other mammals, later feeding on the remains of elk and antelope or seals and whales. As these animals declined in number, condors turned for food to cattle, sheep, coyotes, and deer. The birds became victims of poisoned carcasses set out by ranchers to deter predators, and also died after ingesting lead from bullet-ridden game left by hunters.

In order to control what the newly released condors eat, the recovery team stocks their sanctuary with stillborn dairy calves, and radio transmitters enable biologists to track a bird's every move. Eventually, as the wild population grows, it will become impossible to track each bird, and the condors will fend for themselves.

Following Xewe and Chocuyens, six additional condors were released into the Sespe Condor Sanctuary in 1992, but two subsequently died after collisions with power lines. In 1993, five more condors were freed from those in captivity at the Los Angeles Zoo and San Diego Wild Animal Park, and the breeding program was expanded to include the World Center for Birds of Prey in Boise, Idaho.

At a cost of about $17 million since 1980, the condor recovery effort has shown early signs of success. Its ultimate goal is to restore three separate populations of 150 birds each. But as the condors fly beyond the boundaries of the sanctuary, they will face the same threats they have in the recent past. What's more, available habitat continues to shrink. Time will tell if the California condor can survive on its own.

Cheat Mountain Salamander

Plethodon nettingi

The Monongahela National Forest, on the western slopes of the Allegheny Mountains, is like a fairy-tale forest. It is covered in thick, soft moss which deadens every footfall. On warm, wet, foggy nights, the Cheat Mountain salamanders emerge from under boulders and logs and out of crevices between the rocks, where they have spent the day. These three- to four-inch-long amphibians are foraging for mites, springtails, beetles, and flies. Alert to the slightest vibration, they stand motionless on the fronds of ferns, their dark glistening skin marked with gold or brassy flecks, and their eyes shining.

The Cheat Mountain salamander has no aquatic larval stage, so it does not require standing water. It does, however, require plenty of moisture because it has no lungs and breathes directly through its skin. To facilitate the exchange of oxygen and carbon dioxide, the salamander's skin must stay moist.

By 1920 almost all the old-growth timber in eastern West Virginia had been cut. With it went the overarching forest canopy that the salamander needs to provide shade and inhibit evaporation from its habitat and to prevent dehydration through its skin. For a long time Cheat Mountain salamanders survived in a few isolated pockets of woodland. But even these were threatened by the emergence of second-growth trees, which has caused a resurgence of lumbering activity.

Continued high-elevation coal mining and increasing numbers of ski resorts don't help.

Of the 72 known populations of Cheat Mountain salamander, all are in West Virginia and almost all are scattered within the Monongahela National Forest where the salamander's habitat is protected. These salamanders may have difficulty finding each other for they are wary about exposing themselves to dry environments, especially roads. Scientists are hoping to find niches for at least a hundred self-sustaining populations. Only then can the Cheat Mountain salamander be removed from the list of threatened species.

Chisos Mountain Hedgehog Cactus

Echinocereus chisoensis var. *chisoensis* (= *reichenbachii*)

The decline of the Chisos Mountain hedgehog cactus population may have begun 5,000 years ago with the increasing desertification of what is now the southeastern U.S. More recently, grazing cattle are thought to have impeded the plant's growth, but today the primary threat to the species comes from cactus poachers. It is hard to stumble accidentally anymore upon this three- to six-inch spiny cactus which prefers to grow surrounded by larger shrubs. All the most obvious examples that grew beside roads were taken years ago. Nowadays, the last 100–200 known plants in the U.S. are found on the alluvial flats adjacent to the Chisos Mountains, 10 to 20 miles from the Rio Grande. They all occur within a 30-square-mile area on the eastern side of Big Bend National Park in Brewster County, Texas. Botanists believe there may be many more of these cacti in the largely unexplored regions of the Chihuahuan Desert in northern Mexico.

The most vulnerable time for the Chisos Mountain hedgehog cactus is from April through June when the conspicuous multi-hued flowers appear. Though the blooms do not open widely, the pink to magenta petals, white throat and stamens, and yellow or pink anthers create a striking but short-lived flower. Flowers open around eight-thirty one morning and start to whither by noon. But this is long enough for the pollinators, probably bees, to find it as the cactus

produces plenty of seeds. Unfortunately, few take root.

The Chisos Mountain hedgehog cactus is being successfully cultivated at both the Chihuahuan Desert Research Institute, on the Sul Ross State University campus in Alpine, Texas, and the Desert Botanical Garden in Phoenix. This cactus grows readily from cuttings or through hand-pollination. The scientists at Phoenix, in cooperation with the Center for Plant Conservation, are cultivating as wide a genetic range of plants as they can find, and storing some of the seeds at the National Seed Storage Laboratory in Fort Collins, Colorado.

Cooke's Kokio

Kokia cookei

The progress of the last known Cooke's kokio to grow in the wild was watched by the botanist J. F. Rock, who visited Molokai periodically through the second decade of this century. Before the plant died, Rock transplanted some of the seeds to gardens elsewhere. These plants never developed healthy rootstock, and in the 1930s, with the species facing imminent extinction, a branch of the last *Kokia cookei* was grafted onto the related *K. drynarioides*, itself now listed as endangered.

Since then Cooke's kokio has survived without its own roots. A few further grafts, all clones of the first graft, have been implanted onto other kokio, but, as yet, none have produced viable seeds. To induce the plant to put down roots, botanists have also tried air-layering, wrapping moss around bends in the branch to simulate contact between the branch and the ground, but to no avail. The current technique being attempted is tissue culture: reproducing whole plants from pieces of bud. This has worked with some of Hawaii's other endangered flora, notably the lobeliads.

In the days when Cooke's kokio stood alone, this small tree grew 12 to 15 feet, and produced profuse orange-red flowers each summer. The primary pollinators were probably honeycreepers endemic to the Hawaiian Islands, now almost all extinct.

There were once four species of kokio, one on each of four of Hawaiian large islands, but only three species are left. The kokio are believed to have established themselves soon after each island formed. Though each species became slightly distinct from the others, lacking much competition or many predators, none of them appears to have evolved very much over millions of years. Their chosen habitat is on the drier, western lee side of each island's dominant volcano where, prior to the arrival of people, the chief threat to their habitat was the gradual erosion of the island.

The kokio's rate of disappearance increased as human communities populated the flatter, comparatively dry areas, clearing them for agriculture. The rate rose dramatically with the arrival of cattle, which fed on the succulent leaves and disturbed the soil, making way for opportunistic alien trees and shrubs.

Delta Smelt

Hypomesus transpacificus

This tiny fish may be the force that collars California's rapacious thirst for water. Steel blue with an odor like cucumbers, the three-inch-long Delta smelt lives only in the Sacramento–San Joaquin Delta, which happens to be California's largest source of water for drinking and irrigation.

The smelt inhabits the Delta's null zone, where ocean and river water meet and mix. Larvae that hatch from eggs spawned upstream during the spring spend the summer suspended in this zone, eating zooplankton and growing into juveniles that may disperse into fresh and brackish water. A mature smelt spawns and dies the following winter, completing its life within a year. That makes the fish a harbinger for the annual health of the estuary.

Conditions are best for the smelt when the null zone stays in the shallow, productive waters of Suisun Bay, near the confluence of the Sacramento and San Joaquin rivers. The zone moves, however, under the influences of winter rainfall and freshwater diversions. Delta water was once diverted only during the summer, when most fish aren't breeding, but pumping has increased year-round in the past decade, and except in wet years the smelt can rarely be found in Suisun Bay.

In addition to having their nursery shrink, smelt get sucked into water intake pipes at the pumping stations. Attempts to transfer fish into buckets and return them to the river inadvertently kill large numbers.

Prior to the 1980s, the smelt was one of the estuary's most abundant fish. Since then, populations plummeted to a tiny fraction of their original numbers, though no one knows whether 5,000 or 500,000 smelt survive. Population estimates are made from the number of fish caught during seasonal trawls, and sharp declines have been documented since 1981. It took 12 attempts to net the smelt portrayed here.

Two months after the U.S. Fish and Wildlife Service listed the smelt as threatened, the California Department of Fish and Game followed suit for the state threatened list, although three years earlier the department had rejected such a move. The listings will likely result in more water being left to the benefit of the estuary. While biologists continue to learn what they can about the smelt's habitat, a team at the University of California, Davis, is developing a captive-breeding program to help keep the fish afloat in the Delta.

Devil's Hole Pupfish

Cyprinodon diabolis

Three of the five species of pupfish in the Mojave and Sonoran deserts are listed as threatened or endangered, and all have extremely restricted ranges. The now-extinct Tecopa pupfish (*Cyprinodon nevadensis calidae*) lived only in the outflows of the small North and South Tecopa Hot Springs near Death Valley. It became extinct when the hot springs were converted into baths. The Devil's Hole pupfish lives exclusively in a 10-by-50-foot pool in southern Nevada, probably the most confined habitat of any vertebrate in the world. And in all likelihood, that pool has defined the extent of its range for more than 50,000 years.

Devil's Hole isn't wide, but it is deep. In fact, it was named Devil's Hole because it connects the vast expanse of bright desert terrain above and the underworld aquifer of hot, dark water that lies beneath it. This "skylight to hell" was used by desert travelers in the last century as a place to get water, and perhaps to swim. More recently, it was celebrated as a mysterious and challenging spot for scuba divers to explore an intricate and seemingly endless network of underwater caves, though it is now fenced off. Finally, it has become a sort of shrine to the Devil's Hole pupfish, and to the Endangered Species Act, which came to its rescue.

In the late 1960s when farmers began transforming the desert around Devil's Hole into agricultural land, they drew on the groundwater aquifer to do so. In only a few months of pumping, the water level in Devil's Hole began to drop. As it fell, so did the pupfish's population. It seems Devil's Hole pupfish live on green algae that grows on rock shelves in the hole. These shelves are the only places in Devil's Hole that both remain wet and get enough direct sunlight for the algae to photosynthesize. When pumping made the water level dip below the lip of the highest and largest of the shelves, the algae stopped growing there and the fish lost their primary food source.

A wellspring of concern rose over the fate of the pupfish—and the rest of Ash Meadows, a 37-square-mile oasis that surrounds Devil's Hole and contains at least 26 endemic plants and animals—and the federal government eventually formed the Ash Meadows National Wildlife Refuge and outlawed overtaxing the groundwater that feeds it.

Because of its naturally restricted range and low population, the Devil's Hole pupfish is never likely to be downlisted. To ensure the species' survival in case something should happen to the Devil's Hole population, two captive populations of the fish are being kept in aquariums at Ash Meadows National Wildlife Refuge.

Dwarf Wedge Mussel

Alasmidonta heterodon

A mysterious fish holds the key to understanding the dwarf wedge mussel's reproductive cycle, and perhaps the key to understanding its precipitous decline as well. Female dwarf wedge mussels carry their eggs in their gills, waiting for sperm-laden water to pass through and fertilize them. Once sperm and eggs meet, the tiny larvae are released and seek a host fish to which they cling for a week or two while they metamorphose. Eventually, juvenile mussels drop to the stream bottom where they grow to adulthood.

No one knows which species of fish serves as the mussel's host during that critical early stage of development, but it may be anadromous (one that swims upstream to spawn), since the construction of dams, which interrupt the migration of anadromous fish, has eliminated populations of the mussel. Dams also cause riverbank erosion, because of the constant water fluctuations in dammed rivers, and can leave mussels high and dry when sudden and drastic drops are initiated.

If the unknown species of fish is in decline, that may explain part of the dwarf wedge mussel's problem, too. Conversely, bringing back that fish may be pivotal in accomplishing the mussel's recovery. Lacking a foster parent isn't the mussel's only problem. It is also extremely sensitive to toxic chemicals in agricultural runoff and industrial effluent, which make their way into streams. Siltation and riverbank erosion resulting from dams, livestock grazing, and riverside construction also make life hard for the mussels. In the last 20 years, the number of known populations has dropped from 70 to 19.

Efforts to save the dwarf wedge mussel have become the centerpiece for one Keene, New Hampshire, high school teacher's science curriculum. With authorization from the U.S. Fish and Wildlife Service, he monitors local dwarf wedge mussel populations, and his students learn about river ecosystems by monitoring the mussels' habitat and thus the overall health of the river ecosystem. Students, who test water near the population site for pollutants, dangerously low levels of dissolved oxygen, and other potential dangers, and report their findings to the Fish and Wildlife Service, have adopted the mussel as a kind of mascot and have turned its recovery into a local cause célèbre.

Eastern Indigo Snake

Drymarchon corais couperi

Sometimes growing to eight feet long, the eastern indigo snake is the longest nonvenomous snake north of Mexico and it's arguably the most beautiful as well. Its blue-black scales are polished to an almost glasslike sheen, and when the snake sheds, its abdominal scales gleam with prismatic colors. The indigo snake is extremely docile. It slowly travels great distances in search of food, even through open spaces. Together these qualities make it an easy and attractive target for collectors, who love snakes, and for those who fear them.

Federal and state laws have reduced the trade of eastern indigo snakes, but there is still a black market for them. In Georgia, snake hunters pour gasoline into snake and gopher tortoise burrows, hoping to flush out rattlesnakes to take to rattlesnake roundups, gatherings where rattlesnakes are killed by the thousands and prizes are given for the largest snakes and the most rattlers collected. Because indigo snakes also occupy gopher tortoise burrows, they are incidentally displaced by the gassings.

Even greater threats to the indigo snake's survival are the loss and fragmentation of its remaining habitat. In the northern parts of its territory, the snake relies on gopher tortoise burrows (page 224) for refuge and wintering sites. Fire suppression creates an overgrowth that filters out the sun needed to sustain the undergrowth on which gopher tortoises depend. The decline of the gopher tortoise leaves the indigo snake fewer and fewer options.

Indigos feed with equal enthusiasm on small mammals, birds, frogs, small turtles, toads, lizards, and other snakes. In search of prey, they may range over an area as large as 1,300 acres. As their habitats are invaded by roads, agricultural fields, and residential development, the likelihood that a snake will be run over by a car, caught by a collector, beheaded by a fearful homeowner, or killed by a domestic dog increases, and the species' chance for survival drops.

But there is some hope. In June 1989, the Department of the Interior created a new 30,000-acre wildlife refuge, Florida Panther National Wildlife Refuge, adjacent to the Big Cypress National Preserve. It provides protected habitat for endangered wood storks (page 241), Everglade snail kites, bald eagles (page 217), red-cockaded woodpeckers (page 234), and peregrine falcons, as well as eastern indigo snakes. If more such habitat is protected, and if the public can understand the true nature of this serpent, there may still be time to reverse its slip toward extinction.

Eastern and Western Prairie Fringed Orchids

Platanthera leucophaea
Platanthera praeclara

Orchids are often associated with exotic distant lands. Yet two of the showiest—the closely related eastern and western prairie fringed orchids—were once found throughout much of the midwestern and eastern grasslands. Today, their distribution is almost as broad, but in far, far fewer numbers.

Only recently did botanists distinguish between these two species, doing so on the basis of subtle yet significant biological differences. Curiously, the different structures of the plants' reproductive parts allow the two species to depend for pollination on different parts of the hawkmoth, their primary pollinator. Whereas the western species, with its larger column, deposits pollen, contained within a pollinaria, on the eyes of the visiting hawkmoth, the eastern variety leaves pollen on the insect's proboscis. Both of these exquisite orchids attract their night-flying pollinators with a powerful nocturnal scent and a large nectar reward. With the spread of the tomato and tobacco hornworm hawkmoths (*Manduca sexta* and *M. quinquimaculata*), which have compara-

tively longer proboscises, the western prairie fringed orchid is subject to nectar thievery. These hawkmoths extract nectar without their eyes coming into contact with the orchid's pollinaria, so the flower is robbed of chances of pollination by shorter-tongued hawkmoths.

The prairie fringed orchids, which can grow up to three feet tall, need open, sunny habitat. Flowers only appear in years with adequate rainfall, and these orchids may lie dormant for one or two years. The western species, which has a slightly larger flower and is found primarily west of the Mississippi, has been largely removed from the lowland grasslands and wetlands, now converted for agriculture. Only 1 percent of an estimated 18 million acres of tall-grass prairie remain today. The western species survives in hay meadows which are usually harvested before the seeds have a chance to disperse.

The eastern prairie fringed orchid occurs east of the Mississippi, especially in Illinois and Michigan. In contrast to the western species, the eastern is primarily found in

lowland areas where it is subject to encroachment from other, usually alien species. In centuries past periodic wildfires would burn through the prairie and meadow habitats, clearing them of light-robbing brush. Now such fires are generally suppressed, and much of the eastern prairie fringed orchid's original habitat has been overgrown.

About a quarter of the known prairie fringed orchid sites are protected, and though some public land managers continue to lease orchid habitat for cattle grazing, the populations of both species appear safe for the time being.

Eureka Valley Evening Primrose

Oenothera avita var. *eurekensis*

Sand sings in the Eureka Valley. When the dry, rounded grains cascade down the dunes, a sound akin to a pipe organ or bass viol fills the Mojave Desert air.

Beginning in the 1960s, other sounds began to shatter the peace of Eureka Valley in California. These emanated not from the sand but from legions of dune buggies and off-road vehicles (ORVs) lured to North America's tallest dunes to careen up and down their slopes. They left crisscrossing tracks, torn and trampled plants, and crushed animal burrows. Riders camped and staged their races along the base of the dunes, where the evening primrose grows.

The Eureka Valley evening primrose is a perennial which sprouts a branching stem from the sand. The white flower blooms in the evening, lasts through the night, and wilts with the first rays of morning sun. The primrose is one of three plant and five beetle

species found only in Eureka Valley, an ecological island between Death Valley and Owens Valley. A companion endangered plant, the Eureka Valley dunegrass, may be a survivor from 30 million years ago, when subtropical savannas covered much of the West. The dunes themselves are more ancient than other desert dunes of the West, so the showy evening primrose may have lived here for at least 10,000 years.

In 1976, the Bureau of Land Management closed the dunes to ORVs. Some riders ignored the closed signs and continued to churn up the sand. A year passed before enforcement of the closure brought silence back to the dunes and allowed vegetation to regrow and the scars of tire tracks to heal. Today the threat posed by ORV traffic has largely been eliminated.

Another potential threat is the introduction, by disturbance from nearby road

construction and grazing, of Russian thistle, the ubiquitous tumbleweed, on the eastern side of the dunes where most primroses grow. Both plants are active during the summer, and though the thistle emerges after the primrose peaks, this intruder may be competing with the primrose and the dunegrass for moisture stored in the sand.

The primrose compensates for a relatively short life span and high natural mortality by producing many lightweight seeds that the wind disperses quickly and widely, allowing the plant to colonize new patches of suitable habitat. It now covers hundreds of acres at Eureka Valley, and the population is apparently stable. However, the remoteness and isolation of the region, which may become part of the proposed Death Valley National Park, will warrant keeping the plant on the endangered list.

Fat Pocketbook Pearly Mussel

Potamilus capax

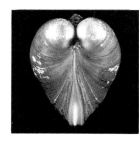

"Five to six miles wide, with crystal clear water that you couldn't dip an oar in without hitting fish and a bottom solid with mussels": that was how Seth Meek described the St. Francis River basin of Arkansas in 1896.

A century later, the St. Francis River system has become one of the most channeled rivers in North America. Where there were once wide expanses of swamp there is now a system of narrow channels and ditches. In most places the water is so dense with silt that a hand dipped beneath the surface disappears.

But in the murky darkness, the fat pocketbook pearly mussel is holding its own. Most of the nine remaining populations are found in the St. Francis Floodway and adjoining streams. The mussel has apparently adapted well to the recently dredged areas where it settles on the sandy, muddy bottoms that do not threaten its thin shell. A few populations exist in the original river that flows parallel to the floodway, but none close to the Huxtable Dam which impounds the lower river.

The fat pocketbook pearly mussel is smoother than most freshwater mussels, with

a shiny, thin, yellowish-to-brown shell lacking obvious growth rings. It grows fast, reaching sizes as big as a man's palm in four to five years, but may not live as long as other, thicker-shelled mussels. Like other freshwater mussels, the females discharge their young from siphons so they can brood in fishes' gills. The young larvae, called glochidia, have evolved in this mussel species to resemble worms so that fish will gobble them up. A few, lucky glochidia pass through a fish's mouth and hook onto its gills. There they may remain attached for several weeks or months, migrating with the fish.

Hitching rides with fish disperses the shellfish, and small populations of pocketbook mussels are found as far away as the Wabash River in Indiana, but this also makes their fate dependent on the health and availability of the host fish. Since adult mussels are filter feeders, they are exceptionally sensitive to environmental contamination; this makes them doubly vulnerable.

Damming, channeling, and agricultural runoff have all contributed to the disappearance of the fat pocketbook pearly mussel

from its known historic ranges. But it has adjusted somewhat in the St. Francis drainage system. Channels that are drained are found to have populations of adult pocketbook mussels when refilled. They either burrow deep into the mud where there is moisture, and estivate, or colonize new areas via the fish host. As mussels go, fat pocketbook pearly mussels go quite well, leaving little trails that look like snake tracks, or the pattern made by someone dragging a finger through the mud.

The U.S. Army Corps of Engineers has recommended that there should be attempts to establish pearly mussel populations on new sites, even ones outside this mussel's historic range. This plan would require a special dispensation under the Endangered Species Act.

Flat-Spired Three-Toothed Land Snail
Triodopsis platysayoides

Though this snail does have a ribbon of teeth for masticating its food, the "tooth" in its name refers to a growth near the opening of the shell. No one is quite sure why it's there, though it likely strengthens the shell, retards dessication of the snail, and hinders potential predators. Moreover, the flat-spired three-toothed land snail has only one such tooth. In fact the unique characteristics of this genus (*Triodopsis*—meaning "three-toothed") are some aspects of its internal genitalia.

Flat-spired three-toothed snails spend most of their time browsing about in leaf litter, the decomposing, loose, moist, cool, dark, deep leaf debris that accumulates around rock outcroppings in the deciduous forest. The snails live among cliffs and boulders where they can retreat into cracks and small niches on hot days. Algae, leaves, and decaying vegetable matter are their primary sources of food. They get the calcium for their shells from soil and rocks and certain calcium-rich leaves.

Most flat-spired three-toothed snails are limited to an area no more than three square miles within the Cheat River Canyon in West Virginia. The largest known population exists in a heavily visited recreational area, and it is the constant trampling of the nearly half-million annual visitors to this park that poses the greatest risk to the snail's survival. The trampling compacts the leaf litter, depriving the habitat of oxygen and moisture, and, in the worst cases, prevents snails from penetrating the debris.

Snails that were transferred for study purposes to a captive colony have reproduced rapidly (since the snails are hermaphrodites they can all give birth). One pair produced 250 offspring within a year, perhaps a quarter of the total estimated wild population. About 150 snails have been reintroduced into their natural habitat, and there are 18 known populations, though most include fewer than 10 adults each. Scientists are finding it difficult to monitor the success of their recovery programs. Too small for radio devices, the released snails have been variously marked with acrylic paint and epoxy, but whatever the marker, other snails eat it from their shells.

Flattened Musk Turtle
Sternotherus depressus

Much flatter than other musk turtles, this five-inch-long, web-footed species with a spine-tipped tail may use its peculiar shape to wedge itself between rocks or to hide in other convenient crevices. A disturbed turtle defends itself with a pungent, musky secretion from glands on the side of its body. It rarely leaves the water and prefers free-flowing streams to sluggish, silt-laden ones—a penchant which contributes to its present plight.

Dams and impoundments built along the Black Warrior River system in Alabama to create lakes have left the turtle's food supply of mollusks smothered by silt. One dam eliminated a section of rapids that may have been a natural barrier between populations of the flattened musk turtle and the stripe-necked musk turtle, which thrives in slow-moving water and may be interbreeding with its rarer relative. Another dam is proposed for the Locust Fork to supply drinking water for Birmingham.

The flattened musk turtle's range coincides with Alabama's major coal region. Strip mines extend to the edges of streams, and toxic runoff pollutes the streambeds where the turtles must find food. Although the mining is subject to regulation, the U.S. Fish and Wildlife Service has never been consulted about avoiding potential threats to the turtle.

In 1985, 30 years after the species was first described, the largest known population suffered from an unknown ulcer-causing disease. As much as half the population in Sipsey Fork, part of a wilderness area and a fairly pristine part of the turtle's range, died that summer from pneumonia and other infections.

A black market in pet turtles poses another threat. Although protected, flattened musk turtles still sometimes appear on price lists for reptile fanciers. Because flattened musk turtles take four to eight years to mature and lay at most a half-dozen eggs each year, recovery, in the face of so many potential threats, will only come slowly.

Florida Key Deer
Odocoileus virginianus clavium

Many Florida Key deer get run over. It's the most common form of death among these deer, killing around 45 animals each year, nearly equal to the annual birthrate. That's what happened to this one too, only she survived. Wildlife biologists hope to be able to release her back to the wild, despite two injured hips and an amputated leg.

Not that there is much "wild" anymore for the Florida Key deer. Once plentiful in the lower Florida Keys, especially those islands where there is permanent fresh water, the deer forage widely on almost all the endemic vegetation, but seem particularly partial to red mangrove, which makes up a quarter of their diet. A subspecies of the Virginia white-tailed deer, the Florida Key deer is the smallest deer in North America, little more than two feet high at the shoulders, with the males weighing about 80 pounds and the females about 15 pounds less. Its ancestors are thought to have been stranded on the Keys after the sea level rose following the last ice age.

Florida Key deer are strong swimmers, and during the wet summers they spread throughout 20 to 27 islands. Each winter they return to two, Big Pine Key and No Name Key, where year-round fresh water is available. The National Key Deer Refuge, centered on Big Pine Key, was established in 1957 after the deer population was estimated to be down to 25–30 animals. Strict protection against poaching brought their numbers back to around 400, but it was about then that the Keys were "discovered." Since 1975 the human population of Big Pine Key has increased fivefold, and the Key deer is quickly being squeezed out. Many of the residents have taken to feeding the deer regularly, and some deer are becoming dependent on handouts. This behavior also draws them to the roadside where they risk being hit by cars and set upon by dogs.

The goal is to establish a stable population of about 300 animals. By lighting controlled burns and eradicating exotic plants, biologists hope to keep the habitat healthy. They are also filling in ditches where mosquitoes breed and where fawns drown, and are digging out new waterholes, but without expanded refuges and an educated public, the chances of maintaining a wild population of Florida Key deer are slim indeed.

Florida Panther
Felis concolor coryi

Cougar, puma, mountain lion, panther. *Felis concolor*—"the cat of one color"—goes by many names. It is hard to imagine that this species of cat was once found throughout the western hemisphere, in almost every habitat: it was the most widely distributed animal, besides *Homo sapiens*, in the Americas. Not anymore. As with any top predator, panthers need to sustain themselves by hunting over wide expanses. For the panther, this can be anywhere from 25 to 250 square miles, depending on the variety and density of its prey. Maintaining a viable population of panthers would require enough land for 100 to 200 animals.

The Florida panther is one of 29 subspecies of panther. Its historic range, which once included much of the southeastern U.S., is now limited to two regions of southern Florida, the Big Cypress Swamp and the Everglades National Park, and its population to a total of no more than 50 animals. Apart from its geographical distinctiveness, this subspecies also has some unusual physical characteristics. The Florida panther is gray-brown, darker and slightly lighter in weight than the other North American subspecies, with longer limbs and smaller feet. Within the limited gene pool, more conspicuous differences have emerged. It has a cowlick along the ridge of its back and a right-angled kink at the end of its tail. More seriously, the males suffer from a condition where only one testicle descends into the scrotum, and more than 90 percent of the sperm is abnormal.

Evidence from DNA sampling indicates that the two populations of Florida panthers have different genetic markers. The panthers found in the Everglades appear to be carrying genes from a South or Central American puma, which mated with captive panthers whose offspring were subsequently reintroduced into the wild. This strain may prove to be the Florida panther's salvation. Because of the dangerously small population and high mortality, particularly among kittens, a captive-breeding program was initiated in 1991. Using in vitro fertilization and artificial insemination, scientists hope to diversify the gene pool still further.

Whether the captive population thrives or not, the wild Florida panther's fate lies entwined with that of its fragile habitat, and it will continue to slip toward extinction unless aggressive steps are taken to protect Florida's wilderness.

Florida Scrub Jay

Aphelocoma coerulescens coerulescens

A distinct subspecies of the widespread scrub jay, the Florida relies on an oak scrub habitat where frequent natural fires keep the oaks under six feet tall. Where pines dominate the scrub, blue jays prevail and the scrub jays disappear.

The scrub jay's diet ranges from larval and adult insects to small reptiles, amphibians, and even baby birds. In the fall, the jays collect acorns and bury them individually in the sand, returning to feed from the cache in winter and spring. A breeding pair defends its 20-acre territory year-round with threatening calls, aerial displays, and fights.

The scrub jay has evolved a fascinating suite of behavior in the isolation of Florida's disappearing oak scrub. Long-term studies at the Archbold Biological Station in Lake Placid, Florida, have revealed the bird's intriguing social system. Pairs mate for life, and juveniles stay with their parents as "helpers" from one to six years. These helpers care for younger siblings, keep watch for predators, and defend the family territory. Parent jays with helpers raise more young than those without helpers. In exchange, the male helpers often inherit the territory or a piece of it, while females retain a safe place from which to locate a mate.

Florida scrub jay numbers have fallen by 80 percent in the past century and continue to drop. Many of the areas that once harbored the jays no longer support them. Three large populations remain on federal land in Florida, including the Ocala National Forest and the Merritt Island National Wildlife Refuge, where management options are being studied. Because the Florida scrub jay's habitat needs to be periodically burned, its future on private land is tenuous and depends on ecologically minded landowners.

Florida Torreya

Torreya taxifolia

In 1875, botanist Asa Gray explored the maze of bluffs and ravines of the Apalachicola River Valley, a unique Floridian habitat where conifers and deciduous trees common to New England abut such subtropical species as palmetto and native bamboo. He came to see "that rarest of trees," the endemic torreya, a conical conifer also called the stinking cedar for the pungent odor of its wood. The tree grew more than 40 feet high in the moist shade beneath taller pines and other hardwoods. European settlers favored the torreya for its rot-resistant wood, which was suitable for shingles, fence posts, and riverboats, but the tree's restricted range limited its use.

The narrow range of the torreya has made the tree all the more vulnerable to an insidious threat. For at least three decades, the population has been stricken by a fungal disease that causes the stems to decay and the needles to drop. Unable to obtain nutrients through photosynthesis, the stunted, skeleton trunk soon dies. Although a new seedling may sprout from a forebear's roots or stump, it, too, succumbs before reaching maturity, which can take 20 years.

A handful of different fungi have been associated with diseased trees, suggesting they are opportunistic invaders of an already weakened tree subjected to some environmental stress. Clear-cutting and replanting on uplands above the torreya's habitat may have altered water or heat levels and reduced the tree's ability to fight infection. An alternative hypothesis posits that smoke from lightning-ignited fires curbed the spread of fungus, which took over after people began to control the natural rate of fires.

So far, no one knows how to curb the disease and the decline.

In Georgia, 27 trees survive on the edge of Lake Seminole. The Florida populations fall on state, city, and private land, all but five trees surviving on the east side of the Apalachicola River. Torreya State Park contains the largest single population. In the wild, some male trees are producing pollen but no female trees are producing seed. So the torreya's best hope lies in cultivating a genetically mixed group of mature, seed-producing trees. For a species whose stock has weathered environmental threats for many millions of years, a future perhaps confined to botanical gardens is a sobering thought.

Furbish Lousewort

Pedicularis furbishiae

In 1880 botanist Kate Furbish came upon this curious plant while documenting the flora of the then-wild Saint John River in northern Maine. Suspecting she had found something new, Furbish sent a specimen to a colleague at Harvard University who verified her hope and named the species "Miss Furbish's wood betony" in her honor. Since then the plant's common name has been abbreviated to the less dignified but still endearing moniker "Furbish lousewort."

By 1976, when the U.S. Army Corps of Engineers launched an environmental impact study of the Saint John River in preparation for the building of a dam there, the plant was presumed extinct. In researching that report, however, botanists found a few small colonies of the plant, and to the chagrin of those invested in the dam, the presence of these seemingly innocuous plants, protected by the Endangered Species Act, brought the project to a halt. Though other major ecological and economic problems contributed to the abandonment of the dam, the meager lousewort became famous (and infamous) for single-leafedly keeping the Saint John the longest free-flowing river in the northeastern U.S.

In addition to its political celebrity, the Furbish lousewort is also an evolutionary curiosity. It is endemic to the Saint John River Valley, but has no known close relatives anywhere in eastern North America. Of the 400 or so members of the genus *Pedicularis*, the vast majority are in Asia, a few in western North America, and two (neither one closely related to the Furbish lousewort) in northeastern North America. If it had evolved along the Saint John, it would likely have close relatives nearby. How then did the Furbish lousewort come to this river valley? Were its seeds carried on winds over the ocean from Asia or across the continent from western North America?

How it got to Saint John is a complete mystery. Why it stays there is less of one. Because the headwaters of the Saint John River are south of its destination, they tend to thaw earlier in the spring than the sections downriver. Water backs up in ice jams which eventually break, sending chunks of ice down the river, clearing its banks of shrubs and trees. No other river in the northeast works quite this way, and the lousewort's success depends on the periodic scouring of the riverside. Without it, competitors would overtake the lousewort's habitat in a matter of years.

Genetic studies of the plant show no variability among individual Furbish louseworts, presumably limiting the plant's adaptability. Over geological time, this genetic limitation exponentially decreases the plant's chances for survival. But there are much more immediate threats. New proposals for hydroelectric dams would, if passed, decrease the total Furbish lousewort population by 60 percent. Upstream logging practices that cause erosion, and thus increase the incidence of flash floods, also pose a threat. Finally, the destruction of lousewort habitat by homeowners who clear their riverside property of vegetation to improve their views is slowly reducing the plant's range.

Gopher Tortoise

Gopherus polyphemus

During the Depression, when they were an important food source for much of the rural Southeast, gopher tortoises were known as "Hoover's chickens." Today, human predation is the least of their problems.

The vast majority of gopher tortoises don't live past their first two years. Many are eaten by snakes or raccoons before they even hatch. In those early years, while their shells are still soft and flexible, they are also vulnerable to predators such as raptors, snakes, and dogs. If a tortoise survives those traumatic years, it stands a good chance of living 40 to 50 years, which is as long as the species has been monitored in the wild.

The female lays a clutch of three to 10 golf-ball-sized eggs and then buries them, often taking advantage of the freshly disturbed soil near the mouth of a burrow. Most of a gopher tortoise's life is spent deep in such burrows. With its powerful, spadelike forelimbs, the gopher tortoise digs long passages that will accommodate its nine-by-six-inch frame. These burrows are regularly over 15 feet long, and may reach 30 feet or even longer, winding around roots and descending 10 to 15 feet underground. A gopher tortoise might dig quite a few such burrows in its lifetime.

Within these burrows, researchers have found up to 360 different species, including 60 different vertebrates. For some of these animals, the burrows are just convenient shelter; for others they are essential to survival. Gopher frogs squat at the mouths of burrows, waiting to capture insects that are attracted by the cool, damp passages. Many reptiles—diamondback rattlesnakes, the endangered eastern indigo snake (page 220), lizards, even baby alligators—are drawn to the burrows for shelter. Bobcats may temporarily move in, and so might the eastern spotted skunk and a variety of rodents. Up to seven species of birds have been seen in the burrows searching for insects or hiding from predators.

The burrows once provided protection from fires which routinely swept through the longleaf pine forests following lightning storms. And not just for the tortoises. Lizards and rabbits have been observed running *toward* a fire and disappearing down a gopher tortoise hole. So upon the gopher tortoise's survival depends the survival of a whole community of creatures.

For the gopher tortoise to thrive, what remains of the longleaf pine forest—about 7 percent of the original extent—must be preserved and properly managed.

Greenback Cutthroat Trout

Oncorhynchus clarki stomias

When settlers first arrived in Colorado, there were four subspecies of cutthroat trout, including the greenback cutthroat which was abundant in the headwaters of the South Platte and Arkansas river systems. But, though they tasted good, cutthroats are small and were not as popular with fishermen. Nor did they take well to the artificial environment of commercial fish farming. So, in the 19th century, more adaptable varieties of trout, such as lake, rainbow, and, especially, brook (brought from Europe), were introduced into the lakes and streams, just as they were throughout most of the West.

By the 1900s, there were hardly any greenbacks to be found, and in 1937 they were reported to be extinct. These reports turned out to be exaggerated, for two small populations totaling about 2,000 fish survived. In 1973, when the Endangered Species Act was passed, the greenback cutthroat trout was one of the first species to be listed as endangered.

Of particular threat to the greenback cutthroat was, and still is, the ubiquitous brook trout. Although as adults the two species usually ignore one another, the juvenile brook trout are aggressive and dislodge the juvenile greenbacks out of their sheltered backwaters into open streams where they are vulnerable to predation, especially when water flow is low. Rainbow trout present another predicament. They sometimes breed with the greenback cutthroat and produce hybrids.

Getting a species officially "listed" can make a difference. After the greenback cutthroat was listed in 1973, brook trout were systematically removed from the two river systems. The greenback population partially recovered, and its status was upgraded to threatened in 1978. For the greenback cutthroat to return to a healthy status, biologists must ensure that the stream habitats remain pollution-free and their ecosystem intact. Permanent physical barriers, however, must be maintained, to make sure that the non-native interlopers downstream do not return.

If the recovery objectives are met, by the year 2000 there should be 20 healthy populations of greenback cutthroat trout.

Green Pitcher Plant

*Sarracenia oreophila
sarraceniaceae*

Like most carnivorous plants, the green pitcher skirts competition from other plants by growing in soils too acidic and wet for other species. High acidity, however, also means low available nutrients, so carnivores glean nitrogenous compounds and other substances from the bodies of their prey, which they lure into their trumpetlike traps with vivid advertisements of the sweet nectar within. Insects are attracted to the pitcher by a colorful hood at the top of each plant. Once inside, they hit a slick patch just below the hood and slide into liquid which has accumulated in the pitcher. Downward-facing hairs inhibit their escape. The plants manufacture their own digestive enzymes, which they use to break down into nutritious juices all but the hard outer skeletons and wings of their prey.

In midsummer, when the seasonal wetland habitats begin to dry, the plant's pitchers, too, begin to dry and wither. In their place are sent up more typical leaves, which persist through spring.

The green pitcher is extinct in Tennessee; one population remains in Georgia, where the species was first discovered in 1875; and the plant persists in a small, and shrinking, number of sites in northeast Alabama. Because of its striking beauty and exotic appetite, the plant is sought out by collectors who have harvested it almost into oblivion. Habitat loss due to agricultural expansion, land drainage, and housing developments also continues to threaten the plant's viability. The supression of natural fires throughout the green pitcher's remaining habitat allows woody plants to outgrow it, and to steal its sunlight and nourishment.

Grizzly Bear

Ursus arctos horribilis

Weighing in at up to 1,700 pounds, the Kodiak grizzly bear, a subspecies of grizzly, is the world's largest terrestrial carnivore, yet often enjoys a nocturnal snack of ripe berries and insects dug from a rotting log. The grizzly bear's fare is as varied as the human diet: roots, fungi, plants, mammals large and small. An adult grizzly can easily dispatch a mature elk with one swat of its massive paw. A particular favorite meal is salmon. Bears that live near salmon rivers may weigh several hundred pounds more than their relatives in more mountainous regions. When the salmon migrate upstream, the bears emerge from their solitary paths and congregate along the rivers, where they threaten and scrap with each other as they establish a hierarchy around the best fishing spots.

By the onset of winter, each grizzly bear may have added up to 400 pounds of fat. It searches out a good protected spot to hibernate—under a rock or in a hollow tree—often using the same den each year. Unlike the black bear, a grizzly rarely hibernates throughout the winter, but surfaces every few weeks. After emerging from their dens in spring, bears roam individually, congregating only during mating season, from May through July. Cubs are born in the winter dens and stay with their mothers for up to two years. Normally shy and wary of humans, grizzlies, especially mothers with cubs, may attack if surprised at close range. Bear encounters can generally be avoided if people move loudly and cautiously through bear territory.

When Lewis and Clark ventured through the West in 1805, upwards of 100,000 grizzlies ranged throughout what would become the lower 48 states. Today, fewer than 1,000 remain. Grizzlies each require up to 1,500 square miles of territory to survive, and have one of the lowest reproductive rates of any terrestrial mammal.

For many people the grizzly bear epitomizes the idea of wilderness. Yet, despite federal protection and growing public support, its future remains doubtful. If this bear is to survive, it will require a commitment from the country as a whole, and especially from the people who live and work alongside the grizzly's traditional domain.

Gulf Sturgeon

Acipenser oxyrhynchus desotoi

Sturgeons have been around a long time. They evolved over 350 million years ago and were among the first bony fishes. Ironically, as they evolved their internal bone structure gave way to cartilage. But sturgeons do have external bones—razor-sharp plates, or scutes, which act as protective armor.

The Gulf sturgeon is a subspecies of the Atlantic sturgeon which was marooned in the Gulf of Mexico after the last ice age. As the water temperature rose, the ancestors of the Gulf sturgeon became isolated west of the Florida peninsula.

Like salmon, Gulf sturgeons are anadromous fish, living in both fresh water and salt water, but unlike most salmon, which return to their natal rivers only once to spawn and die, Gulf sturgeons return every year. With each spring, though, there are fewer and fewer Gulf rivers to which they can return. The sturgeon shown here came from the Mississippi River, but the biggest population, 3,000 or so mature adults, can be found in the Suwannee River in Florida. The Suwannee is relatively unpolluted and is the only major river that flows into the eastern Gulf and is still undammed. Because the bottom-dwelling sturgeons are unable to climb fish ladders, dams and weirs present unsurmountable obstacles to reaching their spawning grounds.

Gulf sturgeons grow rapidly, about a foot each year until they are about eight feet long and sexually mature. They don't appear to eat during their stay in the river, so every winter, back in the sea, they must make up their weight loss.

It was as Gulf sturgeons entered the rivers that, historically, they were at greatest risk, vulnerable to fishermen's nets. The rapid decline of the catch first alerted biologists to the perilous state of the population. Since 1984, fishing for Gulf sturgeons has been illegal, but that decision came too late to save the fish in many of the Gulf's rivers.

A recovery plan is being prepared, but no one knows how many Gulf sturgeons there are, or how many would constitute a viable population. In the meantime, biologists hope that the Suwannee River will remain free-running.

Harperella

Ptilimnium nodosum

The harperella thrives in habitats subject to periodic flooding. This plant has established itself in two separate habitats, one along the shoals and margins of clear, rushing streams, the other on the edges of shallow, intermittently flooded ponds and wet meadows. The harperella's capacity to withstand flooding gives it a strong competitive edge over other, less hardy native plants. However, such a specific ecological requirement means that it has a narrow distribution. Harperella are adversely affected by extreme, prolonged flooding, which simply washes them away, or by inadequate flooding, which gives other plants the chance to move into the harperella's territory.

Various human activities—clearing slopes of vegetation, channeling streams—have led to faster, more violent runoff and excessive sedimentation. Moreover, few of the known sites of the harperella are on federal or state lands and hence subject to protection. In 1984, approximately 10,000 plants were destroyed by the construction of a vacation home subdivision in West Virginia.

Population sizes may vary from a handful of plants in some wetlands to many thousands along certain streams. What constitutes a viable population for this one-foot-high plant is hard to gauge, for its numbers can change enormously from one year to the next.

On the positive side, new populations have been found in Arkansas, Maryland is acquiring more harperella habitat, and landowners are becoming more aware of the precious flora on their soil. The welfare of this plant is being carefully watched. Where the harperella are healthy, so are the stream systems. When they are in trouble, so is their habitat.

Hawaiian Hoary Bat

Lasiurus cinereus semotus

The Hawaiian hoary bat is only one of three subspecies of hoary bats. The other two, the North and South American varieties, are far more numerous than the Hawaiian and, like most other bats, migrate long distances between their summer and winter habitats. The Hawaiian hoary, however, has lost its urge to migrate with the changing seasons. It's just as well, too, as there is no place for thousands of miles around to which it could migrate. So where did the bat come from in the first place? No one knows for sure, though the most likely scenario is that a few migrating mainland hoary bats (or perhaps a single pregnant female) were blown off course by a ferocious storm and came upon Hawaii by sheer luck. If that's how it happened, the journey would be one of the most remarkable mammalian flights of all time, covering at least 2,500 miles.

However it was accomplished, the crossing probably occurred tens of thousands of years ago. In the meantime, the bat has changed, becoming smaller (at about four and a half inches in length) than its mainland cousins and a shade redder.

The Hawaiian hoary is a solitary, tree-roosting creature with a widely scattered population. These adaptations reduce the chance that a large percentage of the population will be killed in one fell swoop, a danger shared by all populations of cave-dwelling bats, such as the Indiana bat and the Virginia big-eared bat (page 239). But Hawaiian hoary bats are hardly impervious to human interference. On Oahu, for instance, where the hoary bat was once common, all of the lowland and most of the upland forests have been destroyed, and the bats are now seldom seen. Throughout the Hawaiian Islands, only about 3 or 4 percent of the native forests remain, the rest having given way to plantations and residential and commercial development. Though no solid baseline data exist on the size of hoary bat populations a hundred years ago, biologists suspect the number has decreased and continues to decline.

Because the bat can successfully forage for insects over the sea, and in sugar plantations or macadamia groves, as well as in its preferred native forest clearings, it probably isn't in danger of starvation. But it is dependent on native rain forests for roosting, and if these forests disappear so will the Hawaiian hoary bat.

Hawaiian Monk Seal

Monachus schauinslandi

This most primitive and tropical of seals belongs to an ancient genus that split into three species: the Hawaiian, Mediterranean, and Caribbean monk seals. The Mediterranean species now numbers fewer than 500 animals, and sometime after 1952 the Caribbean species became extinct. So the future of this ancient lineage may reside with the Hawaiian species, the state's only endemic mammal besides the endangered hoary bat (above).

The largely solitary, sedentary monk seals rest and give birth under the searing sun on coral atolls but spend most of their days and nights at sea, diving deep for fish, lobster, and octopus. The seven-foot-long females may be larger and heavier than males, which do not form the harems commonly seen in other seal species. A high ratio of male to female adults on Kure Atoll and Laysan Island has spurred a bizarre behavior called mobbing. Up to two dozen males gang up on one desirable female, who often dies from injury or infection. Field tests on Laysan Island will investigate whether a testosterone-suppressing drug might temper the males' behavior.

A pup's first year of life after weaning is particularly dangerous, as tiger sharks and gray reef sharks cruise just offshore. To help the pups through this tough time, each year a few small females are flown to Sea Life Park in Honolulu to spend several months fattening up. They are returned to an atoll with a better shot at survival.

The seal's habitat lacks land predators, but human disturbance has taken a tragic toll. As early as 1837, shipwrecked sailors ate monk seals to survive. Then seals were exploited for fur and oil. Entanglements with fishing lines and competition from fishermen for swordfish, bottom fish, and lobsters are among the current threats.

The monk seal still occurs throughout its known historic range but in far fewer numbers. The population apparently plummeted early in this century, then grew for several decades before declining again. In 1982 the beach count of seals was half the number counted 25 years earlier. The population increased throughout the 1980s but dropped after species on which the seals prey declined at French Frigate Shoals. Depending on where one counts, seal numbers are either growing, stable, or in decline.

Hawksbill Sea Turtle

Eretmochelys imbricata

"Lonely Guy" they called it at the Aquarium of the Americas in New Orleans, though it wasn't much more than a year old and was far too young for biologists to determine its sex. This hawksbill sea turtle had been caught in a shrimper's net in a nearby bayou, somehow avoided drowning, and was transferred to the aquarium where for the first three weeks it wouldn't eat. Instead, it gradually rid its digestive system of bits of plastic which it had eaten, as many turtles do in the open ocean, assuming they were jellyfish.

After a couple of years, when the turtle will have a better chance of surviving the man-made intrusions on its habitat and diet, it will be returned to the waters of the Gulf of Mexico to swim again in the warm seas of the tropical areas of the Atlantic Ocean. Its life will still not be an easy one. It must outswim the sharks and continue to avoid entanglement in the shrimpers' nets. And if the latter proves impossible, it must hope that the net incorporates a turtle excluder device (TED) through which it can swim to freedom. If the turtle proves to be female, her dangers will be doubled when she comes ashore to lay her eggs. Hounded by curiosity seekers, and by dogs, and disorientated by beach develop-ments, she may just give up, return to the sea, and release her eggs there, where they have no chance of hatching.

In the 1970s, 50,000 hawksbill turtles were being killed each year for tortoiseshell. On land and at sea, despite international agreements banning trade in products from endangered species, many hawksbills are still prey to the same fate as their ancestors. Their ornate carapaces, which gave rise to the tortoiseshell trade, are still carved into combs and frames for glasses.

Heller's Blazing Star

Liatris helleri

Heller's blazing star is a tough little plant, scrappy and rugged, yet beautiful when in bloom. It has survived for eons on the hostile, wind-swept rocky cliffs in the Blue Ridge Mountains of North Carolina, a habitat that few other plants can tolerate. Unfortunately for the blazing star, humans seek out these rocky cliffs and the winds that rip over them. Hikers walk and climb long distances to view the beautiful Blue Ridge Mountains spread out before them and, incidentally, to stand where Heller's blazing star grows. Unfor-tunately, too, for the otherwise plucky blazing star, trampling is the one form of abuse it can't tolerate.

When the National Park Service inadver-tently put a hiking trail through one of the seven known remaining populations of the blazing star, the plant was almost destroyed, despite park employees' attempts to mitigate their mistake. They built a boardwalk, but people stepped off it to get closer to the cliffs. A ranger specifically assigned to keep people from trampling the plants couldn't be everywhere at once, and the plants continued to suffer. The park service may close the trail altogether, but it is a popular hike, and public resistance to closure would be strong.

Though Heller's blazing star has not yet been analyzed for valuable medicinal com-pounds, two species of the same genus have already been found to contain important anticancer chemicals: an antileukemic drug was derived from *Liatris chapmanii*, and antitumor compounds have been found in *L. pycnostachya*. What could be found in *L. helleri*?

Five of the seven remaining populations of Heller's blazing star are on private prop-erty. State and federal agencies are working with landowners to preserve the populations, but the plant's future is insecure.

Houston Toad

Bufo houstonensis

Lying on the ground next to a dying camp-fire in the rugged south Texas pine forest, sleep is kept at bay by the rustling of uniden-tifiable wild animals, the beat of a raptor's wings, the imagined slithering of viperous snakes. And then, in the background, rises a reassuring chorus of trills, like the tinkling of small, fine bells: the call of the Houston toad.

Bufo houstonensis was probably cut off from the closely related and relatively abundant American toad (*B. americanus charlesmithi*) with the thawing of the last ice age about 10,000 years ago. As temperatures across the continent rose, and humidity fell, their common ancestors were divided and isolated. They have since evolved into the two species we see today, differing in color pat-tern, skeletal morphology, vocalizations, and possibly size.

The Houston toad evolved a dependence on sandy substrates in central and southeast-ern Texas. Unfortunately, the area's growing number of human inhabitants also appreci-ate this habitat and can't resist building on it. Ironically, the creation of two state parks hastened the toad's decline when toad habitat was developed into picnic grounds and recreation areas. Park managers have since become more sensitive to the toad's requirements, though the state is now under pressure to enlarge a golf course, a move that could harm many of the remaining toads.

In the 1950s, the combination of the pell-mell expansion of Houston and an extreme drought led to a radical drop in the number of Houston toads. The U.S. Fish and Wildlife Service listed the species as endangered in 1970. When the Endangered Species Act was passed in 1973, the toad became one of the first amphibians marked for a recovery effort. Although the toad probably once ranged across the entire east-central region of Texas, it is now limited to small populations in only a few counties. A captive-breeding and rein-troduction effort in Colorado County may have restored a lost population there. No conclusive follow-up studies of the reintro-duction project have been conducted yet, and no one is sure whether the toad's num-bers are on the increase or the decline.

The majority of the remaining toads appear to be on private land, much of them on areas near the site of a proposed new airport. If the airport is built, that property will become extremely valuable, and the temptation for property owners to drain and develop it, despite the consequences for the toads, would be powerful.

'Ihi'ihi

Marsilea villosa

Plants are often named after geographical features, but it is rare to find the reverse. 'Ihi'ihilauakea means "wide-leaved and cross-shaped," and it is the full Hawaiian name for a most unusual fern, *Marsilea villosa*, that grows in certain seasonal wetlands. One of those wetlands occurs in a shallow indenta-tion in an extinct volcanic crater in southeastern Oahu—the 'Ihi'ihilauakea Crater.

Although the 'ihi'ihi, as it is more com-monly called, is rare, occurring in only three sites, when the conditions are right, it is the most prevalent plant around the seasonal pools within the crater. Each year, when the rains come, the plants begin to grow from their dormant rhizomes. It is then, as the leaves gradually unfurl, that the plant most resembles a fern. Once the leaves are fully extended, the symmetrical four leaflets could easily be mistaken for clover. Only by looking closely at the leaves is it possible to see the hairs characteristic of the fern family. As the soil dries, hard-walled sporocarps, which carry the male and female spores and require several years to mature, are produced at the base of the plant.

About once a decade, the winter rains are exceptionally heavy, and the area remains flooded for a week or more. The mature sporocarps soak up the water, expand, and open, freeing the two types of sporangia—the immobile megaspores (female) and motile microspores (male)—and allowing for fertil-ization. As the water recedes, the young plant takes root in the moist soil.

The two sites on Oahu where 'ihi'ihi are found are both on protected land. Many of the alien weeds around the vernal pools have been eradicated and much of the off-road vehicle traffic in the area has been effectively stopped by some strategically placed fence posters. A third site was discovered on the west side of Molokai in 1992, and although this land is slated for development, the plant's survival does have the enthusiastic cooperation of the landowner.

226

'Io

Buteo solitarius

The 'io surfs thermals high over Mauna Loa volcano, soaring back and forth, round and round, occasionally emitting a shrill, high-pitched *kee-oh*. But it prefers to hunt perched in a tree, scrutinizing the air and ground beneath it for insects, small birds, mongooses, and small mammals. Until the arrival of humans in Hawaii, the 'io, also known as the Hawaiian hawk, together with the short-eared owl, Hawaii's only other extant raptor, flew atop the food chain, prey to none.

The 'io has the highest degree of reversed sexual dimorphism of any bird in its genus; full-grown males are at least 30 percent smaller than their mates.

Biologists speculate that this discrepancy allows males and females to exploit different foods. The smaller, more maneuverable males hunt insects and birds, while the larger, more powerful females catch mongooses and other mammals. This strategy minimizes competition between mates, allowing them to better exploit the available prey to feed themselves and their offspring. Unlike most mainland hawks, the 'io almost always lays only a single egg, and this one offspring is a big investment of time and energy. It may take an 'io chick as long as nine weeks to fledge, and its postfledgling dependency period may last 18 months.

The hawk, once the adopted symbol of

Hawaii's royalty, is lucky compared with many of Hawaii's birds. It has survived the destruction of much of its historic lowland forest nesting habitat and has adapted to living adjacent to agricultural lands and in disturbed forests on Hawaii, the only island on which it nests. Furthermore, it isn't susceptible to the same devastating introduced-mosquito-borne diseases that have wiped out so many of Hawaii's forest birds.

The 'io's population is thought to be stable, and the U.S. Fish and Wildlife Service has recommended downlisting the species from endangered to threatened status.

Jaguarundi

Felis yagouaroundi

The last wild jaguarundi documented in the United States was a road-kill. That was in 1986, just outside of Brownsville, a town in the southern tip of Texas. There may be more wild jaguarundis north of Mexico, but if there are, they've perfected their evasion tactics. Biologists trying to catch them and provide them with radio-collars have been unable to turn up any.

In Mexico, where the jaguarundi still survives, it occupies the same dense brush habitat as the also elusive, mostly nocturnal ocelot (page 233). The jaguarundi, however, wanders out of the brush and into surrounding areas of dense grassland on its morning and evening hunting expeditions, making it,

theoretically, more visible than the ocelot and an easier subject of research. Theory fails here, however, because although it does emerge from the impenetrable brush during the day, no one has found a trap that can lure the wily cat into its clutches. Ocelots are careful and shy to an extreme, but they seem confident and gregarious compared with jaguarundis.

No one is sure why the jaguarundi is so rare, but certainly the loss of mesquite thickets and other scrub habitat in southern Arizona and Texas is one primary cause. Most of this brushland has been cleared for agriculture and livestock grazing, and what remains is going fast.

Jaguarundis have been successfully bred in captivity, and American zoos are even experiencing a glut of the beautiful cats, but so little is known about their behavior in the wild that putting together a reintroduction program has proven difficult. Though the secretiveness of the jaguarundi has served it well over the millennia, this characteristic may be the final ingredient in its demise. Until biologists can say with some certainty what constitutes sustainable jaguarundi habitat, government agencies will be reluctant to assign rich riparian lands and thinly spread conservation resources to its recovery.

James Spinymussel

Pleurobema collina

Ninety percent of the world's freshwater mussels are found in North America, and at least one species of pearly freshwater mussel is thought to be in every drainage system in the country. They were the basis of the freshwater pearl industry, supplying baroque pearls. Tiny, polished pieces of the inner shell of freshwater mussels have also been found to be the ideal irritant for the formation of cultured pearls in oysters.

Among the 200-plus species within this family of mussels are three spinymussels, including the James spinymussel, which, only 30 years ago, was widespread throughout the James River watershed above the city of Richmond, Virginia.

The James spinymussel lives in the sand and gravel in the bottom of slow-moving, oxygen-rich streams. Placed on the streambed, it will extend its foot and maneu-

ver itself into a vertical position, until it is half buried in the substrate and its gills are facing upward to catch and filter oxygen and food. It is in the gills that the female mussel holds her unfertilized eggs. When sperms, released by a nearby male, come floating by, some find themselves pumped through the female's gills where they fertilize the eggs.

After a few weeks the eggs turn into larvae known as glochidia and the female releases them by the thousands. Within a few days these glochidia must hook to the skin or fins of a host fish or, preferably, be caught in its gills. Different mussel species depend on different fishes. Seven different fish species, all members of the minnow family, are known to host the parasitic stage of the James spinymussel.

About a month later the juvenile mussels fall from the fish onto the streambed and dig

down into the substrate. They will spend their lives—up to 20 years—within a few square feet of their initial resting place. Throughout the country, freshwater mussels have often become the indicator species of the soundness of a river system. Because of the enormous amount of water that filters through their bodies, the health of the mussels reflects the cleanliness of the water and the health of the entire ecosystem.

But with the range of this mussel already drastically reduced, the coup de grace may come from a recent intruder, the Asian clam (*Corbicula fluminea*), which requires no host fish and is more prolific. It seems to prevail wherever these two mollusks vie for territory. Researchers hope to find ways to remove the Asian clam without harming the indigenous spinymussel.

Kauai Hau Kuahiwi

Hibiscadelphus distans

Over 400 Hawaiian plant species are threatened with some level of extinction. There are 12 species with only one remaining individual, 52 with two to 10 individuals, and 165 with fewer than 100. (A further 80 species have recently gone extinct and a further 25 others are probably extinct.)

Although species all over the globe face extinction from habitat destruction, island populations are particularly at risk. More and more native plants must compete with recently introduced plants and animals. These alien species have evolved on continents where there is generally more rigorous competition and now find themselves surrounded by less aggressive species and often with no natural predator.

The fate of the *Hibiscadelphus* (literally, "brother of the hibiscus") genus is typical. Of

six known species, two are extinct, and three are maintained only as cultivated plants. The sixth, endemic to Kauai, *H. distans* (named because of its geographical separation), is actually the most healthy. It was thought to be down to seven plants before a second population of about 100 was discovered on a cliff face in 1991.

Kauai hau kuahiwi is a small tree that reaches 18 feet in height. At about three years, it begins to produce narrow, curved, trumpet-shaped flowers. They are rich in nectar and were probably once pollinated by birds, particularly honeycreepers native to the islands. The extirpation of most of these bird species must have accelerated the plant's precarious position. The Kauai hau kuahiwi in the wild, however, are still producing seeds, though no one is sure how. There

appears to be a low level of self-pollination but that is unlikely to account for the number of seeds. There is also speculation that introduced honeybees are visiting and pollinating Kauai hau kuahiwi.

Today the chief threat comes from feral goats which feed on the foliage, the seedlings, and even the bark. Goat populations are encouraged by the Hawaii Department of Land and Natural Resources for hunting. Thankfully, the animals cannot climb the steep cliffs where most of the Kauai hau kuahiwi still grow. The recovery plan calls for fencing the extant plant populations and eventually eliminating the feral goats. Fortunately, in the meantime, *Hibiscadelphus* is doing well at the Waimea Falls Arboretum and Botanical Garden and the National Tropical Botanical Garden.

Kemp's Ridley Sea Turtle

Lepidochelys kempii

From one to three times each year, on warm summer days, the adult female Kemp's ridley sea turtle crawls ashore to scoop out a hole in the sand and lay her eggs: 100 or so each visit. Almost all the pregnant females come ashore on a 20-mile stretch of beach at Rancho Nuevo in Tamaulipas state along Mexico's east coast. The turtles often arrive at the same time, in aggregations known as *arribadas*. Forty-five years ago, as many as 42,000 turtles came ashore on a single day in June. Today, fewer than 400 make landfall in the entire April to August season.

About two months after the eggs are laid, the hatchlings dig themselves out before dawn and, using the reflection of the rising sun on the water as a guide, scurry past the gauntlet of marauding birds and ghost crabs to the sea. Like many other reptile species, the sex of the sea turtle is determined by the incubation temperature. If Kemp's ridley eggs incubate above 87°F, usually females are born; if below that temperature, most of the hatchlings are male. The females, if they survive, will return to the same beach again in eight to 15 years to lay their eggs. The males will likely never crawl on land again.

The Kemp's ridley is the smallest and most endangered of the seven surviving sea turtle species. At maturity, when its almost circular carapace measures about 30 inches in diameter, it weighs 60 to 90 pounds. By comparison, the leatherback, the largest sea turtle, often weighs over 1,000 pounds.

The life of a young Kemp's ridley was never easy. Mothers have evolved to lay so many eggs because very few hatchlings survive the first two years. Unable to swim fast, and a tasty mouthful for many predators, the small turtles rarely reached maturity. Today, hardly any do. Until Mexico passed a law in 1966 protecting nesting turtles and banning egg collection, few eggs even hatched, and many females on the beach were killed for their shells and leather. Since then the primary cause of death has been shrimp trawlers' nets. Once caught in trawls the turtles were trapped underwater until they suffocated or drowned. Up to 40,000 turtles, including 500 to 5,000 Kemp's ridleys, were thought to die this way every year. Since 1989 turtle excluder devices, (TEDs), have been compulsory on U.S. shrimp trawlers. They are effective, and the turtle death rate has declined substantially. The nesting population of Kemp's ridleys appears to have stabilized.

To circumvent the high mortality of hatchlings, eggs have been incubated, notably at the Padre Island National Seashore in Texas, and reared in captivity at the National Marine Fisheries Service in Galveston, before being introduced into the wild when they are about one year old, giving them a head start. However, not one of the first 23,000 of these released turtles is known to have nested. It is not known why. They may have not yet matured, human manipulations may have interfered with their imprinting to a "home," or natal, beach, or they, too, may have died in shrimpers' nets.

Key Tree Cactus

Cereus robinii

The Florida Keys occur at the junction of the porous Key Largo limestone of Florida and the more compact oolitic limestone of the West Indies. As a result, a distinct habitat has evolved, consisting of species more similar to those of the Caribbean than to those in the rest of Florida. The diversity of plant species in the Keys, which are always limited in distribution, is quickly becoming threatened by the burgeoning human population. The semaphore cactus, *Opuntia spinosissima*, for instance, is down to just nine known specimens, and it isn't even on the endangered species list yet.

This is the only place in the U.S. east of the Rocky Mountains where a large columnar cactus can be found growing wild, but you have to know where to look. The Key tree cactus, once common throughout the Florida Keys, is now reduced to five populations with, at best, a total of 200 individuals. There are thought to be another two populations in Cuba, although there the plant goes by another taxonomic name.

The Key tree cactus is a striking plant, in the true cactus tradition. It typically lives for 100 years or so, growing an inch every couple of years until it is 20 feet or more in height. But these plants are not lone sentinels. They grow in clumps alongside trees and shrubs, their tops level with the ambient canopy. This cactus hammock habitat, or thorn forest as it is known in the West Indies, grows thick and lush in the rainy season.

The long flower of the Key tree cactus is particularly beautiful. It is four inches in diameter, mostly white in color, with touches of pale green and purple. The flowers are even harder to find than the plants, for they bloom only in the evening to attract moths, and each blossom lasts only for a single night. By morning, the flower is already withering.

Whether the Key tree cactus will disappear along with its flowers depends on the vigilance of the administration of the National Key Deer Refuge where most of the extant specimens are found. Their whereabouts is not publicized, for these big cacti are threatened both by those who love them to death—plant collectors—and by vandals who would carve their names into the plant's flesh. Yet the more people understand the cactus's unique value, the greater the likelihood of its survival.

Koloa Maoli

Anas wyvilliana

It is tough to tell the native Hawaiian duck from a mainland mallard. Although smaller and quieter, the koloa maoli resembles a female mallard. In fact, the koloa's immediate ancestor was probably a stray mallard, and the interbreeding of mainland mallards with their island relatives threatens to obscure the koloa's genetic uniqueness.

Mallards were first brought as pets to Hawaii at the turn of the century, but their population has boomed particularly in the past decade. In contrast, the koloa has suffered sharp declines in range and numbers for much of this century. The koloa once inhabited all the main islands except Kahoolawe and Lanai, feeding on snails, insect larvae, and green algae in wetlands and taro fields. Since 1850, sugar cane plantations and residential developments have replaced 30 percent of Hawaii's natural wetlands. Indiscriminate hunting took its toll before 1925, when koloas were protected from hunters. In 1942, a ban on hunting all migratory ducks stopped the inadvertent killing of koloas; however, the population continued to plummet until, by 1960, the species survived only on Kauai. A captive-rearing program began in 1958, with the first birds released on the island of Hawaii the following year. Captive-bred koloas were returned to Oahu in 1968, and when the breeding project ended in 1982, 500 koloas had restocked the two islands. Some 2,000 birds survive on Kauai and constitute the most stable population.

Kauai and Hawaii have high-elevation habitats where koloas have remained relatively isolated from introduced mallards, but most koloas on Oahu are believed to be hybrids. In 1992, Hawaii outlawed the importation of mallards, and plans are afoot to eradicate feral mallards in the islands. Perhaps pure-bred koloas will then continue to make their nuptial flights: a rising female luring one or two hopeful suitors skyward to chase each other in circles before the successful suitor and his future mate come down to earth.

Ko'oloa'ula

Abutilon menziesii

The showy, penny-sized flowers of the ko'oloa'ula bloom year-round and resemble those of its ubiquitous relative, hibiscus. Native Hawaiians may have used ko'oloa'ula flowers in making leis for centuries before scientists first learned about this six-foot-tall shrub when Archibald Menzies collected specimens during George Vancouver's exploration of Hawaii in 1793. Vancouver's voyage also introduced cattle to the islands, thus unleashing one of the plant's major nemeses.

In 1986, only 67 ko'oloa'ula plants in three populations were thought to exist. Additional plants were later found on Lanai, Maui, and Hawaii, and today the species totals around 400 individuals. The largest population, perhaps 375 plants, survives on the island of Lanai. Its habitat largely converted to cropland, the plant occurs today mainly on marginal lands near cultivated pineapple and sugarcane.

In areas where it persists, the plant faces a litany of threats: exotic grazing animals, fire, flood, soil erosion, and development. Cattle nearly eliminated the ko'oloa'ula from the island of Hawaii during one particularly dry year, and feral goats are a current threat to that island's plants. On Lanai, the predominant pests are goats and exotic axis deer. The Chinese rose beetle eats ko'oloa'ula leaves but does not kill the plant.

The ko'oloa'ula takes readily to cultivation in full sun and on well-drained soil. It has been successfully grown at the National Tropical Botanical Garden on Kauai, the Waimea Arboretum on Oahu, and the Amy Greenwell Ethnobotanical Garden on Hawaii. Before wild populations can be restored, the problems of grazing and habitat loss must be solved.

Large-Fruited Sand Verbena

Abronia macrocarpa

There was a time, not long ago, when the fauna and flora of the semidesert were safe from human intrusion. But in the arid, sparse scrub of central-eastern Texas, those days are long gone. First, on one of the two major sites where the large-fruited sand verbena is found, oil was discovered, which brought all the concomitant paraphernalia. Then came the off-road vehicles, followed by a man-made lake, a power station, a residential development with the accompanying need to clear scrub for fire suppression—and, finally, more oil. The other major site is on land used for deer hunting. Though the deer do not destroy the plants, they are thought to eat the flowers. Small wonder the large-fruited sand verbena is having trouble.

The irony is that the large-fruited sand verbena may benefit from disturbance, that is, natural disturbances. All three of the known populations are found on recently stabilized sand, where they took root soon after other vegetation had halted the drift of the blown-out dunes.

The large-fruited sand verbena is described as an herb. During the hot summer the stem dies back, and in the fall a new small rosette develops just above the ground. In the early spring the flower stalk grows up about 20 inches and forms a magnificent rounded head holding 20 to 75 flowers. The flowers open in the late afternoon and close, unless it is overcast, soon after the sun comes up. In the evening they give off an intense fra-

grance, typical of plants pollinated by moths. Recently the pollinators have been identified as several species of hawkmoth. By May, the flowers give way to the distinctively large fruits with their thin, papery walls, ideal for dispersal by the wind.

As all the major populations are on private land and not subject to protection, it is important that the species is cultivated elsewhere. Both Mercer Arboretum in Humble and Southwest Texas State University in San Marcos have a few healthy plants from which biologists hope to propagate viable populations in protected areas.

Laysan Duck

Anas laysanensis

The mallard-sized Laysan duck is by almost all accounts a gentle and trusting creature. The only dissenters are brine flies, Miller moths, and the tiny crustaceans living near the ducks' nests. To these prey, the bird must appear a deadly, night-hunting giant as it stabs its bill into the water with horrible accuracy.

Like many other birds on the Leeward Island chain of Hawaii, the Laysan duck evolved in the absence of threatening predators, so they were unprepared when Japanese feather hunters came in the first decade of this century. Motivated by money to be made in the millinery trade, they killed, in one six-month period in 1909, more than 300,000 birds of various species on Laysan Island, prompting President Theodore Roosevelt to halt the slaughter by forming the Hawaiian

Islands Bird Reservation that same year. More devastating than the feather hunters were the three types of rabbits brought to Laysan by a guano miner who hoped eventually to can and market the rabbits as food. The entrepreneur abandoned his project in 1915, leaving the island to the rabbits.

In the absence of predators, the rabbits multiplied exponentially and within a few years had destroyed virtually all of the vegetation on the two-square-mile island. By the time the rabbits were eliminated in 1923, the Laysan millerbird and the Laysan honeycreeper had become extinct. The Laysan duck was dangling by a thread, with a population low of only 20 birds.

Since then, the island's vegetation has nearly returned, but the duck's population is hovering at about 500 birds. In the 1950s a

captive-breeding program was begun, and Laysan ducks now populate zoos around the world. Because the wild population will always be limited to Laysan Island, and will always be in danger of human-caused or natural disasters, these captive birds serve as an important backup. In 1970 a Japanese ship wrecked on Laysan Island and was feared to have introduced rats, which could have devastated the island's birds by eating their eggs. Fortunately, if there were rats on the boat, they failed to take hold—this time.

MacFarlane's Four-O'clock

Mirabilis macfarlanei

It is not often that a species is named after someone, especially someone who is not a scientist. Ed MacFarlane piloted boats on the Snake River along the Oregon-Idaho border, and it was he who pointed out this striking plant to two botanists who came to survey the region in 1936. The mile-deep Hells Canyon is stunningly beautiful, and its spectacular canyon walls have created a singular warm microclimate, which is probably why this species of the genus *Mirabilis*, along with a few other endemic plants, did not die out thousands of years ago. The MacFarlane's four-o'clock is a remnant from a period when the temperature in the region was much warmer. A second locality was subsequently discovered nearby within a gorge on the Salmon River.

Unlike most members of the four-o'clock family, which close up under the strong sun, the MacFarlane's flowers are open throughout the day. The large, vermilion, trumpet-shaped flowers, each about one inch in diameter, cluster in a flower head of five to seven blossoms. The MacFarlane's four-o'clock is a perennial that grows on the loose, rocky soils along the steep canyon walls. Most sites can only be reached by boat.

Besides the natural threats to the plant—its limited range, the competition from invasive plants such as cheat grass, and the spittle bugs that eat the new growth—there is the increased popularity of the locality, especially since the region was declared a national recreation area. In addition, four-o'clocks are cultivated as ornamental species.

Mirabilis macfarlanei has very showy flowers and is threatened by collection. The U.S. Fish and Wildlife Service has determined that MacFarlane's four-o'clock may be considered recovered when there are a total of 10 protected and managed populations from at least two distinct areas. A population is considered those plants within the daily flight range of a pollinating bee.

Madison Cave Isopod

Antrolana lira

Madison Saltpetre Cave beside the South River, a tributary of the Shenandoah, was the first cave to be mapped in the United States. The mapper was the eclectic Thomas Jefferson. Nevertheless, the Madison Cave isopod was not discovered until 1958, long after the cave had been mined extensively for its saltpeter for gunpowder—and had later become a favorite with spelunkers. The isopod has only been found in three places, two lakes within the cave and in a nearby lake in Steger's Fissure, all within a few hundred yards of each other in Augusta County in northwestern Virginia. This restricted range has caused the isopod to be given a threat-

ened status. The Madison Cave isopod is the only freshwater member of the family Cirolanidae found north of Texas, and it may have survived here since its ancestors invaded fresh water from a marine source perhaps 65 to 70 million years ago.

Less than half an inch long, this subterranean crustacean has no eyes and no pigment in its body. Its life history is still largely a mystery. For instance, although juveniles appear each year, no one has ever found a gravid female, and scientists can only guess its life expectancy. This isopod appears to be omnivorous; both insect remains and organic leaf litter have been found in its gut.

Madison Saltpetre Cave is on privately owned land, but the landlord is cooperating closely with the Cave Conservancy of the Virginias to protect the habitat. In the past the cave has been open to cavers and other visitors coming to see ancient graffiti, including George Washington's signature, carved on the cave wall. Today, the entrance is gated and closed to the general public. Only visitors interested in the welfare of the Madison Cave isopod are given access.

Mead's Milkweed
Asclepias meadii

Mead's milkweed occurs primarily in the loamy to rocky prairies of western Missouri, Iowa, and Kansas, but is also found in the prairies of Illinois. Curiously, these populations appear to be connected by small, isolated occurrences of the plant in a very different kind of habitat, the igneous glades of eastern Missouri. These volcanic areas have relatively shallow soil and are found at much higher elevations than their prairie counterparts. One thing both habitats have in common is the acidic nature of their soils.

The igneous glade populations of Mead's milkweed were described in 1898 but were assumed to have subsequently become extinct. Then in 1991, a botanical survey of the volcanic areas turned up the first of what are now four known sites.

Only about 1 percent of the virgin tall-grass prairie which used to stretch over great expanses of the American Midwest survives. What little remains of unprotected prairie is quickly giving way to housing projects, commercial development, and farms. Hay meadows are the last sizable retreats for Mead's milkweed and these are often mowed when the plants are still in fruit. If the plant isn't allowed to set seed, it can't disperse. The result is a low degree of genetic variation in any one of these populations: most of the plants surviving at a single hay meadow site may be clones. Clones are produced asexually, and are genetically identical to their parents. This does not augur well for the plant's long-term survival. None of its known populations are considered secure.

Missouri Bladderpod
Lesquerella filiformis

As the Missouri bladderpod struggles to maintain a root-hold on the earth, it does have one thing in its favor. One of the largest populations occurs on Bloody Hill, site of a Civil War battle, and now part of Wilson's Creek National Battlefield. Here, it falls under the custody of the National Park Service, which has committed considerable resources to its survival.

The Missouri bladderpod apparently never had a wide distribution; there is no evidence that it ever grew outside a small area of southwestern Missouri. In other ways this bladderpod is particular about its surroundings. It likes open glades and shallow soils over limestone bedrock. It enjoys sunlight but, like most winter annuals, produces its seed and dies before the heat and drought of midsummer. However, the right conditions once met, the Missouri bladderpod can grow prolifically.

In the years after the bladderpod was listed as endangered, annual censuses have shown wide fluctuations in the number of plants in the Bloody Hill glade, from 10,000 to 300,000 in consecutive years. At individual sites the bladderpod may appear in abundance one year, disappear entirely the next, and reappear one or two years later. Although the plant often occurs in stable glades within undisturbed communities of native perennial grasses, it does tolerate small-scale disturbances. Animal activity, drought, or fire can provide windows of opportunity for the Missouri bladderpod to grow abundantly. However, too much commotion could weaken the plant, and it may be this liking for distressed areas that is bringing the bladderpod into direct conflict with exotic plants, notably cheat grass, which also reproduces faster once land has been disrupted.

The Missouri bladderpod is only about eight inches high, with spatula-shaped leaves at its base and narrower leaves along the stem. Its pale yellow flowers appear in late April and give way, after insect pollination, to the "bladderpod," a seed capsule which dries, splits, and opens about four weeks after flowering.

Confronted with attacks from insects and fungi, expansion of the woodland due to fire suppression, growth of the nearby city of Springfield, and the prevalence of exotics, the Missouri bladderpod faces a tenuous future. A comprehensive effort in 1988 did identify 45 previously unknown sites in Missouri, but, as yet, no new populations have been established.

Na'u
Gardenia brighamii

The conspicuous white na'u, or Hawaiian gardenia, flowers and their shiny green leaves were once a common sight on the lower slopes of Hawaii's hills. These 20-foot-high plants with their distinctive, smooth bark were frequently found intermingled with other trees of the forest, and their characteristic, sweet aroma once perfumed the air. Not so now. On Lanai and Molokai na'u were replaced by pineapples, on Oahu and Maui by sugarcane, on Hawaii by pastureland, and throughout all the archipelago by exotic plants and urban sprawl. Today, there are only 15 plants left in the wild, far fewer than the number generally thought to constitute a viable gene pool to offer any hope of long-term survival. Fortunately, hundreds of seedlings are being cultivated in arboretums around Hawaii, and one mature plant has already been transplanted on Lanai. If enough seedlings mature, they will expand the genetic reservoir for the wild plants.

None of the remaining wild plants are on protected federal land. Most are on Lanai where pineapple production is no longer economic and the plantations are fast disappearing. The landowners have arranged a conservation easement with The Nature Conservancy for six small parcels of land, protected from deer and feral animals by eight-foot-high fences. These parcels form the core of the Kanepuu Preserve. The chief threat to the na'u comes from introduced rats which have taken a liking to its fruit, but if environmentalists can work in tandem with the land's owners to establish a healthy reserve, the na'u may last beyond the next century.

Nellie Cory Cactus
Coryphantha minima

When the Nellie cory cactus flowers, it almost doubles its height, but it is still less than two inches tall. The pink to reddish purple flowers appear between March and May, then give way in June to egg-shaped fruits. The flower grows from the top of a thick, tubular, dark green stem which is covered in areoles—soft bumps on the surface—each about one-eighth of an inch apart. From each areole cluster 15 to 27 short, stubby, ashy gray or pinkish spines. Unlike most other cacti, the spines of the Nellie cory thicken away from the stem before ending in a pointed tip.

In the wild, this cactus occurs only in the Chihuahuan Desert scrub in Brewster County, Texas, about 50 miles north of the Mexican border. It lives about 4,000–5,000 feet up in the gravelly, quartz soil. The Nellie cory cactus finds a home in rock crevices on the hard, novaculite outcrops. These cacti have been found on three sites, all within a five-mile radius and all on private land. Their continued survival will require the continuous cooperation of the landowners. Despite their modest size and perhaps in part because of their rarity, the Nellie cory cacti are avidly sought by cactus collectors—a roadside population was quickly extirpated a few years ago. Currently, landowners are resistant to the presence of collectors—and biologists too. There is an unwarranted fear that discovery of an endangered plant on private land will jeopardize the land's value.

Meanwhile, at the Desert Botanical Garden in Phoenix, Arizona, a few plants representing as wide a genetic variation as possible are being hand-pollinated in the hope that a viable population can be maintained.

Nene Goose

Nesochen sandvicensis

When the ancestors of the modern nene goose arrived in Hawaii, they were probably identical to the Canada goose. But in subsequent millennia they slowly adapted to the dry, rocky grasslands and lava flows of their new habitat by shedding half the webbing from their feet. Their longer toes allowed them to scramble adeptly around the hard lava flows on which they preferred to nest. The nene also didn't need the webbing because they seldom swam in their Hawaiian habitat. Since there weren't any predators to flee, or harsh winter snows to migrate away from, their wings began to weaken, and now they fly a lot less than other geese. Young nene take twice as long as Canada geese to develop, a strategy that could only be considered adaptive in a very safe environment. So different has this particularly beautiful goose become that some ornithologists believe it deserves a genus of its own.

There is a paucity of fresh water on the islands occupied by the nene, and the bird relies on fog and dew, combined with a vegetarian diet rich in lush greens, for both nutrition and hydration. When the time comes to nest, nene seek out islands of vegetation, or kipuka, atop lava flows. From a distance it seems a gentle, curious, and trusting bird, but can actually be more quarrelsome and jealous than other species. The gander exhibits especially aggressive territorial behavior, sometimes even attacking his mate.

Biologists estimate that when Capt. James Cooke landed on Hawaii in 1778, there were about 25,000 nene on the islands. But European contact brought a host of problems. For one, thousands of years without predators had made the bird a relatively defenseless, albeit testy, target. Introduced predators such as mongooses and dogs found nene easy prey. So did human hunters and egg collectors. Overgrazing by introduced livestock, such as sheep, also damaged the nene's best nesting and foraging spots. By the 1950s there were only 30 nene geese left.

Since then nene have been bred in captivity with varying success. Today they are raised in several places in the U.S. and Europe, and fledglings are reintroduced to their Hawaiian habitat every year. While still young and flightless, they are placed in predator-proof enclosures stocked with food and water. When the birds are able, they fly over the fence and begin to fend for themselves. The wild population, approximately 500 birds, is still considered extremely vulnerable.

New Mexico Ridgenose Rattlesnake

Crotalus willardi obscurus

The rattlesnake sits curled under a rock ledge, waiting for prey to pass. It hears an approaching intruder via ground-borne vibrations and tenses its solid, muscular body. The snake sees its prey in layered images from its eyes and from the pits on either side of its head that detect heat. Extending a forked tongue, the snake captures the prey's scent before it strikes.

Rattlesnakes have long symbolized the wildness of the West. They attain their greatest diversity in the American Southwest and Mexico. One of five ridgenose rattlesnakes, the New Mexican subspecies is the only North American rattlesnake on the federal endangered list.

It inhabits New Mexico's "bootheel," where flora and fauna from the Sonoran and Chihuahuan deserts converge, together with species from the Rocky Mountains and the Sierra Madre. This is land defended by the Apache until Geronimo's surrender a century ago, mountains from which grizzlies and wolves have disappeared.

The rattler's mottled gray skin matches the pine-needle and oak-leaf litter of its home canyons, where it lives at altitudes of 5,500 to 9,000 feet. The largest known population occupies a single canyon about two miles long and 20 to 200 yards wide. Estimates of its total numbers remain sketchy, because even veteran herpetologists rarely encounter this cryptic species. Its isolation in scattered populations makes *Crotalus willardi obscurus* vulnerable to human and natural threats.

One of the snake's last stands is on the Gray Ranch, an area the size of Rhode Island that contains nearly the entire range of the roadless, untrammeled Animas Mountains. The Nature Conservancy bought the land for $18 million in 1990 and later sold it to a poet-rancher whose private Animas Foundation will manage the land as an experiment in the coexistence of nature and cattle.

Illegal collecting depleted this snake's populations in the past. They can fetch a few hundred dollars in the pet trade, and even the snake portrayed here was confiscated from a collector and turned over to the Museum of Vertebrate Zoology in Berkeley, California. Collecting poses less of a threat now that the Gray Ranch and other areas of the snake's range are protected by private owners. Even so, biologists believe that the New Mexico ridgenose rattlesnake may always remain rare.

Northeastern Beach Tiger Beetle

Cicindela dorsalis dorsalis

Most people rarely get to see endangered species in the wild. Scientists often try to keep their whereabouts quiet. Sometimes, especially where collectors or vandals have shown interest, biologists may be downright secretive. But there are three or four beaches in Maryland where signs invite the public to look, carefully and unobtrusively, and try to see an endangered species in its natural environment. It is the northeastern beach tiger beetle.

Though this beetle was plentiful until quite recently on many northeastern beaches, it was never easy to find. Half an inch long and the color of sand, the adults scurry about the beach around the high tide mark, looking to scavenge on dead fish and invertebrates. Or, if it is July, a male northeastern tiger beetle may be spotted waving his antennae and standing his ground. He is guarding a shallow burrow which a female has recently dug out and where she may be laying her eggs. The burrow is within inches of the high tide line, though laying does not occur during high tide.

When the larvae hatch, they move to the mouth of the burrow, where they anchor themselves in the sand. There, with their heads protruding, they wait for small insects and other arthropods to come within their grasp. That's how they spend their first two years, 85 percent of their lifespan, as larvae. The larvae pupate in late spring and, after two to three weeks, emerge as adults in mid-June—just in time to begin the two-year cycle all over again. By September all the adults have died; only the larvae remain alive to maintain the species.

Accounts written in this century tell of "swarms" of northeastern tiger beetles running around the beaches. But jetties, sea walls, and riprapping of the shoreline have removed much of the beach, and the increasing recreational use of beaches, especially vehicular use, each summer compacts the sand and destroys the burrows. The single Massachusetts population is also the only one surviving on the Atlantic Coast. The other populations are around Chesapeake Bay. As the photographs show, tiger beetles are beautiful—a fact not overlooked by collectors.

Nevertheless, biologists hope to establish at least seven healthy populations distributed throughout the northeastern tiger beetle's historic range, along the coast from Virginia to Massachusetts.

Northern Aplomado Falcon

Falco femoralis septentrionalis

Though only 12 to 17 inches in height, northern aplomado falcons have a wide wingspan, 36 inches, and are exceptionally maneuverable in flight. Sometimes they stoop, falling fast upon the intended victim; other times they may hunt in tandem, flying across their favored grasslands, cutting off their victim's possible escape. If the quarry finds respite in nearby vegetation, the female, the larger of the pair, flushes it out, while the male waits outside the tree stands ready to pounce. The falcon has typical raptor feet, strong with sharp talons. If the prey is small, it may be crushed by these strong claws, but the aplomado falcon generally kills by biting and ripping its catch.

Falcons often appear lethargic, resting during the heat of the day. But they become more active at twilight, when other birds are looking for roosts and bats are foraging. Falcons' keen eyesight can detect animals such as rodents and lizards running in the grass.

Little is known about the breeding biology of the aplomados, but it appears they take part in a courting ritual characteristic of most falcons. It begins with a bowing ceremony between paired birds, and then moves on to "courtship feeding"—with the male offering the female food. Aplomados are probably faithful to a nest site, which generally means they remain with the same partner most or all their lives, renewing the bonding ritual each year. The pair engages in extensive mating and foreplay, sometimes every day from four to eight weeks prior to egg laying. The male provides the female with most of her food through the egg laying, incubation, and early rearing phases.

This falcon nests from January through June, with an average clutch of two or three eggs. The nestlings are dependent on their parents for about 10 weeks. The falcons do not build their own nests but take advantage of old stick nests built by other birds, such as hawks and jays, or use arboreal bromeliads.

Both the northern and the southern ranges of the northern aplomado falcon have shrunk significantly over the last century.

The bird is also extremely rare in northern Mexico, but is still relatively common on the Gulf coastal plain of Mexico in Veracruz, Campeche, Tabasco, and Chiapas. The population of aplomado falcons in the U.S. declined rapidly in the first decades of this century, and the last known nest was seen in 1952. Scientists believe the major cause to be habitat destruction from the encroachment of agriculture. Their final extirpation was believed to be due to intensive agriculture and the use of the pesticide DDT. High levels of DDE, a breakdown product of DDT, have been found in membranes of falcon eggs collected since the 1950s. Pesticide contamination causes the female to lay eggs with thin shells which break easily.

From a captive-breeding program at the World Center for Birds of Prey in Boise, Idaho, northern aplomado falcons have been introduced into the Southwest. There is as yet no sign that any of the released birds are breeding, but the program is only just under way.

Northern Spotted Owl

Strix occidentalis caurina

On silent wings a northern spotted owl glides through a dense stand of Douglas fir in the Pacific Northwest and perches. A rustle in the leaf litter below reaches one of the owl's ears a fraction of a second before reaching the other. The owl homes in on its prey, swoops down, and snatches a woodrat with its talons. It brings the food to its nest site in the jagged top of a dead tree and greets its mate with a contact call—*cooo-weep!*

Unlike most owls, spotted owls have huge home ranges, as large as 6,000 acres. Remnant old-growth forests that have stood for centuries tend to attract the owls. These forests also attract loggers. Hence the dilemma that has made the owl an inadvertent symbol of both its forest home and the perceived gulf between economic progress and environmental protection. It seems as if the fate of the ancient forests, with their intricate web of plants and animals, has come to rest on the owl's wings. Such scrutiny has at least revealed more about the spotted owl's life history than that of most threatened species. As mature forests get cut, the owl must search farther to find food and suitable nest sites in a patchwork habitat. Spotted owls do not build their own nests but borrow abandoned raptor nests or make use of a convenient cavity or platform. Females lay only one to three eggs a year, and nesting success varies widely. Juvenile owls leave their parents in the autumn of their first year and wander until they find a suitable, unoccupied habitat. Given a stable environment, spotted owls live long lives of up to 20 years.

A proposed recovery plan for the northern spotted owl would set aside 196 tracts of habitat totaling 7.5 million acres of federal land. Whether these forest fragments will be preserved and so enable the population to recover remains for the politicians to decide.

Northern Swift Fox

Vulpes velox hebes

The northern swift fox has earned the dubious distinction of being protected where it isn't found and not protected where it is. The smallest North American fox, just the size of a house cat, it weighs a mere five pounds. Its short, stout limbs and low body help it hit a speed of 25 miles per hour while pursuing jackrabbits. Naturalist Thomas Say, who described the swift fox in 1823, likened it to "a bird skimming the surface of the earth." The nocturnal fox also hunts rodents, birds, lizards, and insects, covering a territory of four to 17 square miles. During daylight, the fox stays close enough to its burrow to escape from its predators—coyotes, bobcats, eagles, and owls.

Although swift, this fox is less wary than other canids. A willingness to eat meat baited with poison made it an unintended victim of the war waged against wolves and coyotes. The fox was also trapped for its fur, only to have the small, coarse pelts dyed to resemble more highly prized mammals. In the quarter cen-tury after 1853, the Hudson's Bay Company purchased more than 117,000 swift fox pelts.

Today the northern swift fox suffers from both scarcity and ambiguity. It has disappeared from all the northern border states, and only 50 to 75 persist in South Dakota. In a one-year period swift fox numbers in that state's Buffalo Gap National Park apparently fell from 40 to five. The ambiguity concerns whether the northern swift fox is really distinct from the southern subspecies, *Vulpes velox velox*, which, though more common, is also declining and is still trapped for fur.

The last hope for the northern swift fox may lie in a Canadian captive-breeding and reintroduction effort begun in 1971. Swift foxes imported from as far as Colorado formed a core breeding population that now numbers 39. Since 1983, after an absence of nearly half a century, some 500 foxes have been released in remnant prairie. Maybe half of these have survived, offering hope that this fleet-footed predator, once a symbol for a sacred society of Alberta's Peigan tribe, will continue to grace northern grasslands.

Oahu Tree Snails

Achatinella spp.

There is only one case where a whole genus has been listed as an endangered species. It is the genus *Achatinella*, the Oahu tree snails. Of 41 species of Oahu tree snail, 22 have already recently become extinct, and of the remaining 19, all but two are on the verge of extinction, and some may no longer be found in the wild.

Oahu tree snails come in an assortment of beautiful colors—red, orange, yellow, brown, green, gray—and an extensive variety of patterns. They are divided into three subgenera based on their shell form, which can be spherical, conical, or egg-shaped.

Achatinella means "little agate," and these snails have been called the "gems of the forest." It was their beauty that caused their first major threat. They evolved on Oahu for thousands, possibly millions of years in balance with their predators. Then humans arrived. Early Polynesians used to string these half-inch to one-inch gems into anklets, bracelets, and necklaces. Then, between 1850 to 1900, came "land shell fever," when collecting tree snails was a craze among Europeans and Americans, some of whom possessed over 100,000 shells. By the 1930s most species were becoming rare, and some were already extinct.

Tree snails are, as one might expect, sedentary animals. They even mature slowly, taking four or five years, one-half of their life span, to reach sexual maturity, and then only producing about four offspring per adult each year. They often spend their lives in a single tree, almost invariably an indigenous species, which makes them especially vulnerable to forest destruction. Eighty-five percent of Oahu's native forests have been lost. If a snail's home is cut down, it is unable to adapt to another species of tree.

Most of Oahu's remaining forest is a conservation zone, but the tree snails still must survive what may be the most effective predator of all. In 1955, Hawaii's Department of Agriculture, looking to eliminate a highly destructive African snail introduced into Hawaii by a gardener 20 years earlier, brought a carnivorous snail, *Euglandina rosea*, from the mainland. The result, for *Achatinella*, has been disaster, with some tree snail species driven to extinction within a year.

To combat these multiple threats, scientists are trying to develop a toxic "snail burger" which will eliminate the introduced species. They are also rearing the tree snails in captivity and intend to return them to areas inaccessible to the carnivorous snail.

Ocelot

Felis pardalis

Ocelots are so shy, so reclusive, spending their nights hunting and their days hidden in thick brush, that no one, even the biologists devoted to their recovery, knows much about their habits in the wild. What they do know, what the world agrees on, is that the ocelot is an irresistibly beautiful animal, so alluring that in the 1980s European furriers bought ocelot skins for as much as $4,000 apiece. Even today, there is a flourishing market for skins from South America, where hunters can still find cats to hunt.

There are too few ocelots left in the U.S. for hunting to threaten the animals—besides, listing under the Endangered Species Act makes killing an ocelot a federal crime punishable by a year in prison and hefty fines. Habitat loss and fragmentation continue to stalk the cat, however, as the few remaining areas of dense brushland (so dense that crawling through it on hands and knees is almost impossible) are transformed into farms or consumed by creeping urban sprawl.

In the early 1990s the first analyses of ocelot scat revealed the staples of the cat's diet: rodents, cottontail rabbits, and birds. Ocelots require large territories to hunt and breed, and few patches of brushland left in south Texas are large enough to support communities of the cats. Land on which such dense brush grows tends to be quite fertile, and thus attractive to farmers.

Conservation biologists do have hopes for an experimental program run by the U.S. Fish and Wildlife Service and Texas A&I University, which will restore ocelot habitat on 250 acres of reclaimed agricultural land in the Rio Grande Valley. But restoration is no substitute for preserving the few areas of remaining wild ocelot habitat. Re-creating habitat is always difficult, and growing the dense foliage the ocelot needs will take decades.

Pharaoh, the tame male ocelot photographed for this book, was bred in captivity and raised on a private retreat in Sonoma County, California, far from its native home.

Piping Plover

Charadrius melodus

When a male piping plover is ready to mate, he flies figure eights and gives a piping call in an attempt to seduce a female into his territory. This accomplished, he lands, puffs up his feathers, struts, and peeps and whistles around the female while thrusting out his neck and stomping his tiny feet. The nonchalant female doesn't appear to pay much attention, but shows her approval by allowing the male to mate with her.

Mated pairs incubate their clutch of four eggs in shallow scrapes in the beach sand, which they sometimes line with shell fragments. While one parent sits on the eggs, the other feeds, but never beyond calling distance. If a predator appears, the bird on the nest can summon its mate and together they will attack the predator or try to divert it from the eggs by pretending to have a broken wing and limping away from the nest.

Within hours after hatching, chicks begin darting about hunting for the tiny insects and invertebrates that will be their staple diet for the rest of their lives. Studies have shown one-day-old chicks to wander as far as a mile, in the company of their parents, in their search for food.

Most piping plovers nest on the shores of the Atlantic Ocean or above inland rivers in the Great Plains, but a few raise their families on the shores of the Great Lakes. All piping plovers migrate, wintering in the south either on the Gulf Coast from Florida to northern Mexico or on the Atlantic Coast from North Carolina as far south as the West Indies.

The six-and-a-half-inch-long birds were nearly hunted to extinction early in this century, before passage of the Migratory Bird Treaty Act in 1918 led to a resurgence in their numbers by the late 1920s. But the birds came under renewed attack, this time by less intentional but harder to regulate threats: waterfront development and recreation. In addition to the roads and homes built atop plover nesting habitat, beach buggies and other off-road vehicles crisscross the birds' habitat and destroy their nests. Pet dogs brought to the beach also terrorize plovers, separating parents from young and occasionally even catching and killing chicks. Garbage brought to beach sites by growing numbers of visitors has attracted foxes, skunks, opossums, and rats, which also relish the tiny landbound birds. Piping plovers nesting in the Great Plains were hit by the many water projects that diverted major rivers there.

Because the piping plover is listed as endangered by Canada's Committee on the Status of Endangered Wildlife, this agency and the U.S. Fish and Wildlife Service are cooperating on many aspects of its recovery plan. They are trying to minimize disturbance by fencing off nesting areas from off-road vehicles and predators and by enforcing pet leash laws. The Great Plains population is steadily decreasing despite these efforts, and the piping plovers nesting in the Great Lakes area are holding steady at a perilous 20 pairs. But the Atlantic Coast population is showing some signs of recovery.

Pitcher's Thistle

Cirsium pitcheri

Beginning 14,000 years ago, ice age glaciers retreated north from the basins of the Great Lakes, and a series of sand dunes took their place. Some time during this period the Pitcher's thistle migrated east from the Great Plains and found a new home in the stabilized lakeshore dunes. The thistle relies on heavy waves or wind to shift sand landward and create the right conditions for it to colonize new habitat by releasing seeds to the wind, each carried by a tiny parachute, called a pappus. After five to eight years as a juvenile with a low profile, the thistle flowers and sets seed between May and September. It sends up a three-foot-high stalk supporting as many as 125 flowering heads. Each head has at least 20 tiny flowers, which contrast with the woolly stems and spine-tipped leaves.

Because a thistle population naturally declines as other vegetation moves in and fills the dune habitat, biologists concern themselves with metapopulations, or a mosaic of loosely interacting subpopulations. Any short-lived subpopulation can come and go with the vagaries of environmental variables, but the larger metapopulation persists—provided enough seeds disperse into enough suitable habitat. Inadvertent building on dunes that lack Pitcher's thistle at a given time can still fragment and disrupt these metapopulations.

Thistle habitat has been disappearing beneath buildings in both Michigan and Wisconsin, and construction of marinas and condominiums continues on Lake Michigan and Lake Huron, especially around port towns and river mouths. Sometimes, dunes get bulldozed to provide lake views for homeowners, and unwitting residents pull up the thistle, thinking it is a weed.

In addition to the residential threats are the recreational ones. Hikers and hang gliders inadvertently trample plants. A 1987 survey on Manitoulin Island, Ontario, found crushed plants within off-road vehicle tracks at four out of five sites. Planned roads will bring more people to the shores of the Great Lakes. Finally, sand mining and the building of jetties and retaining walls to control lake levels impact the supply of sand and affect the factors that create suitable thistle habitat.

Currently, Pitcher's thistle ranges from the north shore of Lake Superior south to Indiana and east to Georgian Bay, Ontario. It is classified as threatened in Canada, where it occurs at a dozen sites. At the state level, the thistle is listed as endangered in Wisconsin, and threatened in Michigan and Indiana, where it occurs, and in Illinois, where it has not been seen for more than 70 years. Michigan harbors nearly 90 percent of the remaining U.S. populations.

Plymouth Redbelly Turtle

Pseudemys rubiventris bangsii

Most freshwater turtles love to bask, and the redbelly turtle is no exception. In fact, basking is crucial to the turtle's well-being. It elevates internal body temperature, assists in digestion and shedding, and can discourage leech infestations. On any mild day redbelly turtles haul themselves out of ponds in Plymouth County, a few miles east of Boston. But to find these turtles it is important to know where to look, for only about 15 ponds support this species, and most of them are located in an area smaller than three square miles.

Other populations of redbelly turtles are found in fairly large numbers along the coastal plain from New Jersey to North Carolina. But, in Plymouth County, redbelly turtles became isolated from their southern cousins several hundred years ago. The Plymouth redbelly turtle has a domed, dark mahogany shell with colorful reddish orange markings. Probably because of the cool winter climate in Massachusetts, it experiences a longer hibernation than its southern counterparts, so its annual activity period is shorter and it must nest a few weeks later, usually in late June or July. The consequence of this nesting delay is that many hatchlings do not survive the succeeding winter.

In 1986, Massasoit National Wildlife Refuge, a permanent federal wildlife refuge for redbelly turtles, was established. It includes most of the Massachusetts ponds known to be currently inhabited by the Plymouth redbelly turtle. There is no public access. Turtles are at maximum risk from predators—raccoons and skunks ashore, bullfrogs, fish, and wading birds in the water—while they are young. So each summer researchers protect the nests with wire-mesh cages and keep a portion of the hatchlings in captivity over the winter. These "head starts" grow rapidly, and by the time they are released the following spring, they are too large for the jaws of their enemies.

Presidio Manzanita

*Arctostaphylos
pungens* var. *ravenii*

Westerners are familiar with the tall, treelike manzanitas that pepper the California chaparral with their beautiful ruddy and muscular branches and peeling bark. But more than a hundred different plants fall under the name *manzanita*, and many don't resemble these shrubs at all.

For instance, the Presidio manzanita grows only about one foot off the ground, and covers an area about 14 feet by 12 feet, a profile adopted against the plant's wind-blasted coastal habitat in the San Francisco Bay region. It is an easy plant to overlook, but close examination reveals delicate white blossoms tinged with pink, and bright red

berries hidden beneath the foliage. The small serpentine outcrop in the Presidio Army Base where the only native Presidio manzanita grows is an extremely unusual and, for most plants, inhospitable habitat. But this unique manzanita does well here in this patch of serpentine soil, high in magnesium and low in calcium, and so do the few other rare native grasses and annuals that grow up protected within its reaches.

The Presidio manzanita is a special case among endangered species. As far as botanists know, it has always been extremely rare, limited to an area only five miles wide, and occurring infrequently even there. So its

recovery plan contrasts markedly with that of species once more widely distributed. The objective is to protect the one wild plant known and to cultivate seedlings and cuttings to keep the plant's genetic lineage alive when the solitary Presidio individual dies.

Price's Potato Bean

Apios priceana

A 15-foot vine, the Price's potato bean produces a bean, which is inedible, and an underground tuber, like a potato, which is edible. The tubers of a related plant supplied protein for Native Americans, and Price's potato bean may one day serve as a genetic reservoir providing disease resistance for this relative or may become a food crop of its own.

Since Price's potato bean was discovered in Kentucky in 1896, 36 populations have been found in five states. Of these, only 25 exist today, and the plant has apparently disappeared altogether from Illinois. Most of the remaining populations occur on private property and lack protection; almost two-thirds of them possess fewer than 30 plants each. The potato bean does not bloom every

year, and the tuber can survive dormant in the ground, so botanists hope that more populations will be discovered.

In fact, almost all aspects of the plant's life history remain a mystery. No one knows how seeds for new vines get dispersed and what conditions permit them to grow. Honeybees and bumblebees and a single butterfly species pollinate the potato bean's flowers, but why the plant produces few seeds is one of many unanswered questions. Growing the plants in greenhouses will help botanists learn more about the plant and its needs, and seeds from captive plants may restore wild populations.

Because it prefers forest edges and sunlit openings, the potato bean actually benefits from light logging in its habitat, but clear-

cutting threatens populations. Conversely, the natural closing of the forest canopy allows more competitive, shade-tolerant plants to crowd out the potato bean. In the future, wild potato bean populations may need to be maintained by selective logging or by weeding, mowing, or spraying other plants. The U.S. Fish and Wildlife Service won't consider the potato bean safe until at least 40 protected populations can survive on their own for a decade.

Red-Cockaded Woodpecker

Picoides borealis

The red-cockaded woodpecker is only one of hundreds of species that depend on the Southeast's shrinking old-growth pine forests. But it is so conspicuous, and so fascinating, that it has become the fulcrum of the debate over the future of the pine forests, much as the northern spotted owl is the center of attention on the opposite corner of the continent (page 232). Like the spotted owl, the red-cockaded woodpecker does not simply prefer old-growth forests—it requires them for survival. As the once-great forests have been ground into pulp, so the bird's numbers have plummeted.

The red-cockaded woodpecker's social and material life revolves around the cavities it excavates in pine trees that are on average between 80 and 100 years old. Trees this age are often infected with red heart disease, a

nonlethal fungus that softens the heartwood and allows woodpeckers to dig their cavities. Tall, older trees are also safe sanctuaries from the fires that used to frequent the woodpecker's habitat, and from shrub-climbing predators, most notably the rat snake. Red-cockaded woodpeckers make small holes, called resin wells, above and below their nesting cavities. Whereas old trees ooze resin from these wells at just the right pace—forming a sticky sort of moat around the cavity opening which keeps snakes at bay—young trees ooze too much resin too fast, sometimes covering the cavity entrance and locking birds out or, worse, trapping them inside.

Cavities sometimes take years to complete and may be used for decades, handed down from fathers to sons. Each "household"—

centered around the male's nesting tree cavity—usually includes a breeding pair, their offspring of the year, and a couple of "helpers"—male offspring from previous years. The helpers contribute to incubating the eggs, brooding and feeding the young, maintaining resin wells, and excavating new cavities. Young females, on the other hand, usually take off in late winter and wander through the population until each is accepted into a clan that needs a breeding female.

Advocates for the birds are trying to restrict the clear-cutting of old-growth pine forests, but intense lobbying by the lumber industry and the federal agencies that oversee logging have kept the trees flowing to the pulp mills and the birds heading toward oblivion.

Red Wolf

Canis rufus

How fickle humans are, pursuing the wolf deep into forests, high onto mountains, poisoning, shooting, trapping, trying to rid it from the face of the earth. And, with the red wolf, almost succeeding. Then, missing its call, its wildness, its freedom, humans doggedly apply themselves to its return.

Bringing the wolf back, however, is more difficult than was hunting it down. By 1980 all known wild red wolves were captured and enrolled in a breeding program, and the animal was declared "biologically extinct" in the wild. Since then progress has been made. Red wolves are now being born and raised in zoos and research institutions throughout

North America. In 1987, four pairs of adult red wolves, born and reared in captivity, were reintroduced to a wildlife refuge in northeastern North Carolina—the first reintroduction of a predatory mammal ever. Today, approximately 25 wild red wolves roam free.

The red wolf is a shy creature, traveling in groups of only two or three, hunting alone, and staying clear of humans and other large animals. It is most active at night, when it hunts rabbits and other small game. It mates in January and February, and has litters of between two and eight pups in the spring.

Biologists who study the red wolf are

unclear just how pure the animal's genetic lineage is. As the number of wild red wolves shrank, the incidence of interbreeding with coyotes increased. No one knows for sure how long such interbreeding has gone on, or just how much it has influenced the red wolf's genetic makeup. Some researchers suggest that the red wolf may even be a hybrid species, the product of mating between gray wolves and coyotes. Others believe it is a subspecies of gray wolf. Still others say its distinctive build and coloration indicate it is a species unto itself.

Running Buffalo Clover

Trifolium stoloniferum
(Fabaceae)

When the running buffalo clover was first considered for the endangered list, no one knew if the species still existed. Only two occurrences of wild running buffalo clover had been recorded since 1910. Then, in 1983, a botanist working for The Nature Conservancy discovered two plants in West Virginia. They were found alongside an off-road-vehicle trail in a forest. By the following year, vehicles using the trail had killed or severely damaged several clover plants spawned by the two parents.

This was a sad yet hopeful chapter in the history of a plant that once dominated Kentucky's bluegrass region and which may have evolved in an intimate association with the once vast herds of buffalo. Perhaps buffalo benefited the clover by cropping other, more competitive plants or by creating disturbed areas that the clover could exploit. Some sightings of the plant have occurred on former buffalo migration routes. And one West Virginia population persists in that state's last known buffalo range. The 1816 diary of explorer and botanist David Thomas led researchers to the site of two surviving clover patches in Indiana.

Today there are approximately two dozen known populations of varying size, and the U.S. Fish and Wildlife Service hopes to find or establish 30 stable sites on protected land throughout the plant's range. Right now, most populations are not producing seed. Unfortunately, the historic sites have been so altered that scientists don't really know what the right conditions are for running buffalo clover, or even whether such habitats still exist.

Santa Cruz Long-Toed Salamander

Ambystoma macrodactylum
croceum

If you stand near one of the Santa Cruz long-toed salamander's breeding ponds on just the right rainy night in late September or October, you will see the ground begin to squirm. Salamanders will emerge from animal burrows, from spaces between and along the roots of plants, right out of penetrable dirt. Others, already aboveground, will march with maniacal determination toward the road from as far as a mile away.

The days of sex that follow the fall migration are hot and heavy for these long-toed amphibians. Once the females have mated, however, they are in no mood to linger, and with the next rainy night, they leave their eggs behind and head for the hills, or the woods. The males, on the other hand, get to the ponds a little earlier and hang around a little longer than the females, presumably hoping for extra sexual encounters.

About eight days after females have fixed their eggs (about 200 for every female) to stalks of spike rush or other submerged plants in the ponds, and the ponds have been abandoned by most adults who have returned to their summer feeding grounds, the eggs hatch and the creatures' metamorphoses begin. Over the next 90 to 140 days, they change from tadpolelike larvae to fully formed, land-dwelling salamanders. By the end of the summer, when the breeding ponds have dried up, the first-year salamanders seek moisture by burrowing into the pond site. With the first winter rains, when the older salamanders are migrating toward the pond, the juveniles flee the site of the upcoming orgy, not to return for another year, when they are sexually mature.

Not one of the nine known populations of the Santa Cruz long-toed salamander is safe. Even the two that are located on state and federal reserves are vulnerable to pond siltation and drought. In the last few years, interagency cooperation has come a long way in protecting the salamander, though, and in 1992 the population at the Santa Cruz Long-Toed Salamander Ecological Reserve was given three tunnels to help the salamanders pass underneath a road that runs between some of the winter and summer habitats. The tunnels, it is hoped, will keep the migrating salamanders safe from by far their most dangerous threat: humans in a hurry.

Schaus Swallowtail Butterfly

Heraclides aristodemus
ponceanus

In anticipation of mating, the female Schaus swallowtail butterfly rests on the ground, raises her abdomen, and stretches out her trembling wings—which may measure five inches from tip to tip—so that they are nearly flat to the ground. The male hovers above and behind her for a long moment before making contact.

Her eggs fertilized, the female makes her way to a torchwood tree or a wild lime, where she lays her eggs on the upper sides of leaves. Four days later the eggs hatch into caterpillars, which continue to occupy the same trees, eating the tender new leaves. In about 20 days, the caterpillar attaches itself to a branch and molts into a stiff chrysalis. The larva is covered with a mottled tan, yellow, and white pattern and closely resembles a bird dropping. But if this camouflage doesn't suffice and predators probe the larva, a hydraulic device, called an osmeterium, emerges from a slit at the front of the thorax and releases a flow of repulsive, odoriferous fluid.

The pupae remain dormant usually for one year, but sometimes for several years, before the adult butterfly emerges for its final, airborne stage of life. Most adults live only a few days and the longest-lived last only three weeks.

Most of the butterfly's original habitat was lost to development early in the century. Spraying for mosquito control further decimated the population. By the 1970s it survived only in Key Biscayne National Park and on the northern part of Key Largo. This limited habitat was thought to be secure, then Hurricane Andrew hit Florida in August 1992, passing right through the butterfly's population center. All of the butterflies, which were in their pupal state, were submerged beneath several feet of salt water for hours. Much of the torchwood and wild lime habitat on which the creatures depend was also destroyed.

Serendipitously, researchers from the University of Florida had begun to conduct a captive-breeding program just two months before the disaster hit. The butterfly in the photograph was part of this captive population and had just emerged from its pupa when the picture was taken.

The 1993 census found about 100 butterflies in Key Biscayne National Park. That was enough to impress researchers who had feared there would be none. But it wasn't nearly enough to secure the population's future. The captive population may hold the only hope for the future of this species, a hope that is now as delicate and vulnerable as the butterflies themselves.

Scrub Mint

Dicerandra frutescens

For thousands of years, four species of perennial mint in central Florida—longspurred, Lakela's, Christman's, and scrub—have survived, even thrived, although each is restricted to a discrete habitat and a small geographical range. Within its range, each mint could be found only on the unshaded, bare sand at the edge of sand pine scrub vegetation. But the expansion of human settlements and, in the case of the Christman's and scrub mints, citrus groves has reduced all four mints to a few small sites.

The scrub mint, however, has found an ideal habitat. It grows successfully in the bare sand of several fire lanes that cut through Archbold Biological Station near Lake Placid, Florida. Its only other sites, all within a few miles of the town of Lake Placid, are less hospitable. Two sites, for example, are right in the middle of a proposed residential development. In one case the streets have already been laid out, though most of the houses have yet to be built. The provisions of the Endangered Species Act do not apply to plants found on private land.

Despite the scrub mint's idiosyncratic requirements, once the right conditions are met, the plant grows easily. The release of pollen from its anthers is triggered by bee flies. Although the scrub mint has no defenses against human expansion, it has evolved distinctive abilities to defend against more traditional enemies. When insects chew on its leaves, the plant releases a powerful chemical that smells like peppermint oil. A small whiff of this repellent sends ants and other insects scurrying. Even more remarkable is a caterpillar which eats the scrub mint leaves and then regurgitates upon itself to deter its natural enemies. Scientists hope that a synthetic version of the chemical will be useful in commercial insect repellents.

If the scrub mint, with its unique properties, is to survive, federal agencies must acquire more land where additional populations can thrive unmolested, and private landowners on whose land the scrub mint is already growing will need to acknowledge that the plant has a value beyond economics.

Sensitive Joint Vetch

Aeschynomene virginica

It took a boat trip at high tide through the shallow waters near the mouth of Chesapeake Bay to find this particular example of the sensitive joint vetch. The plant lives in a distinctive habitat: within a narrow band of riverine marshes that are far enough from the bay to contain essentially fresh water, yet are still subject to the bay's flooding tides. A number of other plant species have adapted to become dependent on these unique, fresh, tidal wetlands and they are candidates for the endangered species list, but the sensitive joint vetch is the first to be federally listed.

The name *Aeschynomene* is derived from the Greek for "bashful." Pliny the Elder, in discussing medicinal plants, mentioned "the herb aeschynomene, so-called from the shrinking of its leaves at the approach of the hand." Botanists know that the distinctive closing of the leaves is caused by rapid changes in water pressure, but they still don't understand how this reaction benefits the plant. Ironically, despite its common name, this vetch is less sensitive than many others and has to be uprooted before it shows the characteristic folding of the leaves.

Although the sensitive joint vetch is an annual, it has sturdy branches and can grow as high as six feet before it flowers in July, producing striking blossoms. Like many legumes, the sensitive joint vetch may contribute to the supply of soil nitrogen by producing root nodules in symbiosis with nitrogen-fixing bacteria.

This joint vetch and its habitat are facing numerous threats: water projects that may divert water and increase salinity; dredging and channelization; and highway development, including plans for an outer beltway around Washington, D.C. With the vetch's addition to the endangered species list, some of these proposed plans may require modification to ensure the continued existence of this rare plant and the ecosystem that sustains it.

Shortnose Sturgeon

Acipenser brevirostrum

Known by a host of vernacular names, including pinkster and roundnoser, the shortnose sturgeon can be distinguished from the more common Atlantic sturgeon by its shorter, rounder snout and an overall size of only three feet. The largely nocturnal shortnose probes stream bottoms with its snout, tastes with its whiskerlike barbels, and vacuums up food along with mud and gravel that it spits from its gills. Juveniles eat crustaceans and insect larvae, while the adults take mollusks, worms, and shrimp.

Compared with its ocean-going Atlantic relative, the shortnose sturgeon stays close to fresh water. Once every three to five years, between January and May, females, having fasted for months, leave salty, brackish bays and estuaries for fast-flowing, freshwater streams. A female can lay up to 200,000 tiny, brown eggs, which the males fertilize as the eggs sink and stick to rocks and weeds. A small percentage hatch and fewer still survive, but those that do may live more than 50 years.

On several rivers, dams have impeded the sturgeon from returning to spawning grounds. Two landlocked populations now survive year-round in fresh water after dams blocked their return to the sea. Soil eroding into rivers smothers sturgeon eggs. Spawning sturgeon must also run a gauntlet of pollution from mills and chemical plants along the lower reaches of several rivers.

The shortnose sturgeon's historical range stretched from the Saint John River in New Brunswick to the St. Johns River and estuary in Florida. It probably occupied every large river running to the Atlantic but has disappeared from many. Stable populations persist in Canada and New England and in South Carolina and Georgia, and large numbers inhabit New York's Hudson River and Maine's Kennebec River. The Saint John River harbors the last Canadian shortnose population, which is threatened by pollution from pulp mills and insecticides. Though several rivers contain more sturgeon than scientists previously thought, the shortnose has many rivers to repopulate before it is out of danger.

Small Whorled Pogonia

Isotria medeoloides

The small whorled pogonia is one of the rarest orchids in North America. Even if one knows just where in the eastern deciduous forests to look, it is not easy to find. For one thing, individual plants tend to grow separately, rather than in large, conspicuous colonies. For another, its white blossoms are tiny and not at all outstanding from any distance. Viewed close up, however, the five-leaved whorl and the delicate blossoms are beautiful and quite distinctive.

At least twice in the twentieth century, the small whorled pogonia has been a victim of botanical fads. Early in the century, before there was much understanding or concern about the importance of preserving plant populations in the wild, the rarity of the plant brought botanists from far and wide to see, and to collect, it. The few known sites of small whorled pogonia were trampled and dug up by enthusiastic botanists in search of a glimpse, and often a specimen, of something so rare.

More recently, the booming wildflower fad has brought amateur botanists and flower collectors into the woods looking for new and different wildflowers. In both Connecticut and New Hampshire, collectors and possibly vandals raided the areas where the endangered pogonia is found, doing what may have been irrevocable damage to those populations. The raiders probably won't be able to transplant their booty successfully since terrestrial orchids such as the small whorled pogonia are dependent on very specific soil conditions for their survival. For instance, the plant must grow in the presence of a particular soil fungus, but no one yet knows what fungus that is—it may even be an unknown species.

Education and enforcement can reduce collection, but the small whorled pogonia faces other, more formidable challenges. Residential and commercial development poses major threats to the pogonia's remaining habitat. In Virginia, two colonies are on still undeveloped lots in a quickly growing housing subdivision, one colony is on land slated for development, and a fourth colony grows on a strip of land along a highway.

The U.S. Fish and Wildlife Service has conducted extensive searches for new populations of the small whorled pogonia, tripling known populations in 10 years. It has also made successful efforts to protect remaining populations, such as the largest one in New Hampshire, which, after a popular landowner campaign, achieved protected status.

Steller's Sea Lion

Eumetopias jubatus

The first Western naturalist to visit the northern Pacific Ocean was George Steller, the surgeon aboard the Russian ship *Vitus Bering*. While shipwrecked on Bering Island in 1741, he described two enormous marine mammals he found. One was the slow-moving Steller's sea cow, a sirenian like the West Indian manatee (page 239), which may have weighed up to 10 tons. It made excellent eating for the visiting seal hunters, and within 30 years, it was extinct. The other, the Steller's sea lion, survived, but in the last few years its population has entered a rapid decline.

The Steller's, or northern, sea lion is by far the largest of the eared seals—fur seals and sea lions. The largest males may weigh over 2,000 pounds, three times the weight of the average female. After the males haul out on their rookeries in early May, they stake out a territory—the best are closest to the water—which they defend against other males. They rebuff the challenges, hissing and snorting, lunging with their upper bodies and occasionally drawing blood. For the next 60 days they will not risk leaving their realm and must survive on their stored reserves. During this period the females arrive to bear and wean their pups.

The females come into estrus about two weeks after giving birth, and if a male is to pass on his genes, he must be constantly vigilant since females appear to show no attachments to a particular male, even within a single season. Despite the bulls' best efforts, females clamber between male territories almost at will.

Steller's sea lions have been protected under the Marine Mammal Protection Act since its passage in 1972. This act, forbidding the deliberate killing without a permit of all marine mammals, ended the trade in the sea lion's fur. Nevertheless, the population has continued to decline, especially in the last few years, and on April 4, 1990, the species was put on the threatened species list. The chief culprit appears to be commercial fishing. Fishermen compete for the sea lions' prey species, notably the preponderant walleye pollock, which has been in greater commercial demand as other fish stocks have declined. Sea lions also drown when trapped in fishing gear, and occasionally they are deliberately shot by frustrated fishermen.

Stephens' Kangaroo Rat

Dipodomys stephensi

The Stephens' kangaroo rat is well adapted for evading its predators. Its enlarged inner-ear cavities help it pick up low-frequency sounds and distinguish, say, the scrape of a snake's belly against the ground from the beat of an owl's wings overhead. If the predator is a swooping owl, the kangaroo rat hops casually to the side where, because an attacking owl has little lateral control, it is out of harm's way. If the danger is a rattler, it jumps backward, beyond the snake's striking distance.

But when the threat is human development, it knows no route of escape. Not long ago, western Riverside County, in southern California, east of Los Angeles, was continuous lowland grassland, interrupted intermittently by small hills and granitic outcrops. Flat, well-drained grasslands are the kangaroo rat's preferred habitat. Unfortunately, they are also the best lands for farming. As agriculture has divided most of the area into a mosaic of farmland, the kangaroo rat has been relegated mostly to patchily distributed granitic outcroppings and hills, where food sources are scarce and populations genetically isolated.

The Stephens' kangaroo rat is a part of a complex web of predator and prey species that once ranged throughout the region: deer mice, cactus mice, wood rats, cottontail rabbits and blacktailed jackrabbits, bobcats, coyotes, weasels, diamondback rattlers, coachwhip snakes, striped racers, granite spiny lizards, orange whiptails, side blotched lizards, great horned owls, burrowing owls, red-tailed hawks, and American kestrels, to name only a portion. Conservation that protects the Stephens' kangaroo rat and its habitat will also protect many of these species.

The U.S. Fish and Wildlife Service is trying to pull together, through land trades and purchases, small parcels of kangaroo rat habitat into larger tracts on which their numbers can multiply and their genes disperse. Some conservationists are skeptical, however, that without a freeze on the continuing loss of remaining habitat and a vigorous restoration program creating new habitat and linking it with old, the kangaroo rat's chances are slim.

Stock Island Snail

Orthalicus reses reses

Stock Island snails are excellent reproducers, capable of exponential population growth in short periods of time. Like many mollusks, they are hermaphroditic; each individual contains both female and male reproductive parts. Though they can't fertilize themselves, a pair can do the favor each for the other. One, starting out as the male, strokes the female with outstretched eye stalks till she exposes her pore, then injects his sperm. The next day they may swap roles, the former male receiving the fluid transfer.

It is a sad irony that such a quickly reproducing snail is, as far as anyone knows, extinct in the wild. Once crawling about the trees in the dense, junglelike hammocks of Key West in south Florida and its small neighbor, Stock Island, the snail was driven from most of its range by real estate development and the introduction of the fire ant, a species able to locate and devour the snail's eggs and newly hatched young. In June 1992, only three snails of the entire original population were left, all of them adults and all living in an ant-infested area beside the Stock Island golf course.

About 10 years earlier, the U.S Fish and Wildlife Service had transplanted a few Stock Island snails onto the grounds of an old mainland hammock, now an amusement park called Monkey Jungle, in southern Dade County, Florida, where they flourished in the native tropical hardwood trees dotting the car park.

When Hurricane Andrew ripped through Florida in August 1992, Monkey Jungle was in its path. The foliage canopy under which the snails normally took daily sanctuary was torn off the trees, exposing the little refugees to the hot, dehydrating sun. If a University of Florida researcher and his students hadn't arrived on the scene shortly after the hurricane, the snail would probably be extinct today. They collected all the living snails they could find, a total of 463, and brought them to the University in Gainesville to be bred there in captivity. The U.S. Fish and Wildlife Service contributed 200 terrariums to the effort. If all goes well, there could be as many as 20,000 snails by 1994. The trick will be finding a suitable place to reintroduce the snails, whose native keys can no longer accommodate them.

Swamp Pink

Helonias bullata

In April, when winter's taupe hues are still turning green and before many wetland plants have flowered, the precocious swamp pink sends up a hollow flower stalk as high as two feet off the ground and opens its pom-pon of tiny pink or purplish flowers. For uncountable millennia, these fertile flowers have augured the arrival of spring, the season of fecundity in bogs, meadows, and swamps along the eastern seaboard from South Carolina to southern New York state.

Despite its confident posture and irreverent blossoms, the swamp pink is a vulnerable plant. Subtle changes in the hydrology of a wetland may destroy the conditions needed for its survival. It is often found alongside skunk cabbage, cinnamon fern, red maple, Atlantic white cedar, swamp azalea, sweet bay magnolia, and sweet pepperbush. But unlike these hardier plants, the swamp pink is dependent on very specific, and little understood, hydrologic conditions.

This poses a serious conservation dilemma. The alteration of property adjoining, but not actually on, a wetland area may still have a deadly impact on a swamp pink population. If the development project is above the plant's location, for instance, it may alter the hydrologic conditions only slightly, but still enough to destroy the swamp pinks growing below. Since federal and state agencies don't have the authority to regulate the development of lands adjoining protected areas, conservationists often can do little to protect those unlucky populations of the swamp pink.

There are still small and vulnerable populations of the swamp pink in Delaware, Maryland, Virginia, North and South Carolina, and Georgia. New Jersey, with 30 populations of the plant, is the swamp pink's last real stronghold. Though some of the New Jersey populations are protected in the Pinelands National Reserve, the largest and most viable ones are on private land where they are threatened by residential development.

Tennessee Purple Coneflower

Echinacea tennesseensis

In a red cedar glade about 14 miles outside of Nashville, Tennessee, the extremes of temperature, moisture, and light are too much for most plants, but they are ideal for an incredible array of 42 specially adapted or endemic species. One of these endemics, the Tennessee purple coneflower, once thrived in such glades throughout the state. Now it is extremely rare, and existing populations are feeling the heat of Nashville's expanding city limits.

The Tennessee coneflower faces another threat—one that is less tangible, but more insidious. Historically, the plant grew in protected isolation from the other species in its genus, separated from them by both geographical and biological barriers. In recent years these barriers, whatever they were—no one is sure—have been broken down, perhaps by the introduction of new species of pollinating insects or changes in the distribution of native pollinators, or perhaps by the migration corridors made by the roads that have penetrated the area. One known contributor is the intentional planting of attractive species related to the Tennessee purple coneflower along nearby roads. Whatever the causes, there is no question that the once-distinct plant is in danger of being cross-pollinated by other species of its genus and is losing it unique qualities.

A nonendangered relative of the purple coneflower, *Echinacea purpurea*, has an old and widespread reputation as a powerful herbal remedy. Before the popularization of antibiotics in the 1920s, *E. purpurea* was the most widely used herbal remedy in North America.

Recent scientific studies in Germany have shown that compounds in *E. purpurea* may help people fight colds and other viral infections, and a growing demand in the U.S. for a tincture made from the plant has sent some enthusiastic amateur herbalists into the field to collect their own. Unfortunately, they sometimes collect the endangered *E. tennesseensis*, whose medicinal value is untested, confusing it for *E. purpurea*. With a plant this rare, a few mistakes can tax significantly the plant's already slim chance for long-term survival.

Most of the Tennessee purple coneflowers are on private property, and their survival depends entirely on landowners' willingness to spare them. Three colonies do survive on state lands, however, where the Tennessee Division of Forestry has agreed to protect them.

Texas Blind Salamander

Typhlomolge rathbuni

In the warm natural aquifers beneath the city of San Marcos in Hays County, Texas, lives a five-inch-long creature few people have ever seen. It is the Texas blind salamander. The salamander has no functioning eyes, though just beneath the skin are vestigial eyes, just two black dots. The skin of this troglodytic (cave-dwelling) animal lacks pigment, but it is full of neuromasts, highly sensitive nerves that pick up the slightest vibration.

This salamander is highly adapted to its unusual environment. It has thin sinewy legs and a long spatula-shaped snout. It moves slowly as it crawls through the tiny tunnels that crisscross the aquifer, waiting to sense its prey—minute snails, copepods, shrimp, and other crustaceans that enter the aquifer from the surface. These minuscule creatures are most plentiful after a heavy rain, which brings them down through the substrate.

A few captive specimens of the Texas blind salamander are in the Cincinnati Zoo, but little is known about this salamander other than its extremely limited distribution. In the 1960s its unique characteristics attracted scientists and collectors to the few places in its subterranean habitat that were accessible to humans, especially Ezell's Cave in Hays County. When the cave was sealed to keep people out, it also kept out bats whose guano was probably an important source of the salamander's nutrients. There is no recovery plan for the Texas blind salamander because there is no evidence that the population is declining. If it were, there would be nothing anyone could do to help reintroduce the salamanders into such an inaccessible environment. Their status depends entirely on a stable aquifer filled with unpolluted water. As the city of San Marcos grows and continues to draw water from the aquifer, the well-being of the Texas blind salamander will represent a crucial monitor of the health of the habitat beneath the city.

Thick-Billed Parrot

Rhynchopsitta pachyrhyncha

No sight or sound captures the spirit of the Arizona wilderness like a flock of screeching thick-billed parrots speeding at 60 miles-per-hour over the state's mountain pine forests and vast valleys. The birds' piercing screeches are so loud they can be heard from two to three miles away. The parrot, one of only two native to the continental U.S., still lives in the highland forests of northern and central Mexico, but was wiped out in Arizona early in this century by miners, loggers, and soldiers who hunted it and destroyed the forests on which it depended.

Like all parrots, the thick-billed is a very social bird, living in tight-knit groups that roost and travel together. Their conspicuous presence, trust in humans, and easily locatable screech made them easy targets for hungry hunters. The Arizona population, once encompassing flocks of hundreds of birds, used to be continuous with the Mexican population through highland pine forests, which the birds use for food and nesting. Parrots would follow the availability of their favored pine nuts up and down this long forest corridor.

But much of that forestland has been cut, and though Mexico's parrot population still numbers in the thousands, the corridor has been broken and there is no way for the birds to move north into Arizona. No way, that is, except the smuggler's basket. The birds, which can fetch between $250 and $750 on the U.S. black market, are often smuggled into the country and occasionally are discovered and confiscated at the border. Ironically, those confiscated Mexican birds have formed the core of the species' restoration program in Arizona.

Between 1986 and 1989, 50 thick-billed parrots were reintroduced to the Arizona backcountry. The program was showing promise, and there was at least one instance of successful wild breeding. But in 1990 the new Arizona population suffered serious setbacks from fires, drought, attacks by predators, and "social disintegration." When Arizona Game and Fish reintroduced the birds into the wild, wild-caught birds were first put together with naive, captive-reared birds to beef up the population and to encourage the wild-caught birds to "tutor" the naive ones. Even for parrots, however, a little knowledge can do a lot of harm, and the social fabric of the mixed flock came apart at the seams. The captive-raised birds took off from the main group and became open to attack from goshawks and other predators.

Early in 1993, 28 wild-caught parrots were reintroduced to the Chiricahua Mountains, and they showed signs of being better prepared for the wild than their predecessors. The flock was seen to withstand its first hawk attack by circling in the air and squawking as the red-tailed hawk approached. Captive-reared parrots will be reintroduced into the flock slowly, one bird at a time, so they don't disrupt the flock's social organization.

Tooth Cave Spider

Neoleptoneta myopica

It spends its life in the darkness of 40-by-80-foot Tooth Cave and a few nearby limestone caverns, hiding under rocks and ready to capture passing insects. Though the scientific name of the Tooth Cave spider implies nearsightedness, its rudimentary eyes have lost their retinas, leaving the spider blind. Merely a tenth of an inch long, the spider is a third the size of an endangered pseudoscorpion that also lives in Tooth Cave.

Tooth Cave is part of the Jollyville Plateau, a limestone karst region northwest of Austin, Texas, in Travis County, that has long been home to a unique suite of invertebrates. In addition to the spider and the pseudoscorpion, this cave harbors an endangered ground beetle, mold beetle, and harvestman. Once thought to occur only in Tooth Cave, the spider has since been found in three other caves on the plateau.

One of these sites, New Comanche Trail Cave, has been infested by fire ants, South American natives that spread into Texas after being inadvertently introduced to Alabama. Fire ants found the extreme humidity and 70°F temperatures inside the caves to their liking. They proceeded to strip the caves clean of life: spiders, scorpions, salamanders, and all.

Nothing can eradicate the ants, but vigilant control measures—pouring boiling water on ant mounds or baiting them with a growth-inhibiting chemical mixed with cornmeal—have held them off, a hundred feet from the gated entrance to Tooth Cave. When one ant colony dies, however, another takes its place. And any leftover bait must be retrieved before nightfall, when cave crickets come out and forage.

These crickets are a critical part of the food web in the caves. The Tooth Cave spider may eat newly hatched crickets as well as other invertebrates that feed on fungi growing on cricket droppings. Other aspects of its life history remain hidden in the darkness of the caves.

A more pressing threat than even the encroaching fire ants is the spread of development from Austin. Caves south of the city have been filled, paved, and buried beneath houses, buildings, streets, and parking lots. Now Austin's city limits have crept north, threatening Tooth Cave and others in the vicinity.

Texas Earth First! catalyzed an effort to strike a compromise under an amendment to the Endangered Species Act that encourages habitat conservation plans. This compromise measure, called the Balcones Canyonlands Conservation Plan, would permit some development while setting aside enough habitat for the spider and the other listed cave invertebrates to survive. A 890,000-acre area would protect habitat for the endangered black-capped vireo and golden-cheeked warbler, as well as two rare plants. After much tension and frustration for all involved, the plan is almost complete, and if approved by the U.S. Fish and Wildlife Service, it would secure all known habitat in Travis County for all but one of the endangered cave invertebrates.

Virginia Big-Eared Bat

Plecotus townsendii virginianus

A Virginia big-eared bat comes into the world absolutely hairless, its face masked by long, limp ears draped over still-unopened eyes. In a couple of days its ears perk up, its eyes open, and it begins to grow hair. At four weeks it starts to fly, and within six weeks is fully weaned of mother's milk and old enough to leave the cave each night on summer hunting expeditions.

Until then, it is left behind to hang alone in the recesses of its warm summer caves while its mother takes off early in the evening to hunt for moths and other insects in the nearby forests and fields. Although the mother may return to the cave up to three times during the night to nurse her young, as the offspring grows older it may have to wait patiently for its mother while she takes shelter, often in an abandoned building, for a midnight nap. After digesting food captured early in the evening, the female feeds again before finding her way back to the cave. At about sunrise, the young bat's waiting is rewarded by its mother's return and her nourishing milk.

In winter, the colony occupies a different cave altogether—or at least a different section of the cave—and only a very cold, undisturbed cave where the temperature remains steady will do. There they hang, in dense clusters, their heart rates and respiration slowed almost to a standstill, their temperatures dropping as low as 40°F. And so it has been in these caves, probably for countless millennia.

This ancient way of life is a delicate one. Because a single cave may support thousands of bats, even a seemingly minor disruption can have devastating consequences, not just for the bats living there, but for the entire species—over half of the entire population winters in a single cave in West Virginia. During the 1960s and 1970s, when caving became popular, bat populations declined.

Vandals and uninformed cavers have taken a large toll on the Virginia big-eared bat: in 1988 the population of one colony dropped from over 1,000 bats to fewer than 300 in one year as a result of vandalism in their cave.

The bat's recovery plan includes the construction of cave entrance gates and fences, penetrable by bats but not people. The plan also calls for further study of the mysterious creatures. The extent of their habitat, for instance, is still unknown. No one is sure just how far they need to fly from their caves to find nourishment.

The cave ecosystems in which the bats live are also little understood, but they are known to include isopods, amphipods, spiders, millipedes, mites, and beetles. Many of these creatures depend on the bats to contribute nutrients, in the form of dead bats and guano, to their otherwise rather sterile cave ecosystem.

Walker's Manioc

Manihot walkerae

The one accessible wild Walker's manioc in the U.S. grows in the lower Rio Grande Valley where it produces flowers each spring and fall, shortly after the seasonal rainfall. The flowers, some male, the rest female, appear in clusters of three or four. They open in the late afternoon and are gone within a day. During those few hours, they are visited by a variety of potential pollinators—bees, wasps, ants—yet not many of the flowers actually set seed. And, for the last few years, despite intensive efforts, none of the seeds has germinated.

At the University of Texas at the Austin Rare Plant Study Center, biologists successfully reproduced the plant both from rootstock and by tissue layering, incubating the seeds in covered dishes at 80°F.

From observing the larger population of this plant in Mexico, botanists have deduced that Walker's manioc seems to prefer the open sunlight, but grows most successfully surrounded by native scrub. In the U.S. almost all this brushland has been cleared by mechanical and chemical methods to create land for agriculture and residential development.

The Walker's manioc is a close relative of the cassava *Manihot esculenta*, which is a food staple on three continents and throughout much of the tropical Pacific. Cassava crops are being threatened by a number of bacterial diseases, and scientists hope that by introducing genes from related species, cassava will become more resilient. Despite its apparent inability to maintain a healthy population of its own, Walker's manioc may make a significant contribution to the cassava's genetic enhancement—provided, of course, that the rare manioc is not allowed to go extinct.

In the meantime, the Walker's manioc will also require some genetic infusion of its own from cross-pollination between the Mexican and the U.S. stock. Then, when 15 populations, each with at least 100 plants, have been established, the plant will be considered recovered. That target seems light-years away.

Welsh's Milkweed

Asclepias welshii

Few species can survive in the aeolian—wind-driven—sand dunes where Welsh's milkweed lives. This herb grows each year from a thickened rhizome in both stabilized and drifting sand. It rises as a cluster of stems and reaches between 10 and 40 inches tall before putting out cream-colored flowers each spring. These flowers attract a variety of pollinators: bees, wasps, butterflies, and moths. But the Welsh's milkweed can also reproduce vegetatively—in fact, that may well be its primary method of reproduction. It appears to favor the lee side of the dunes where it sprouts from rhizomes, which keeps it just ahead of any advancing sand.

With such a variety of reproductive methods, the Welsh's milkweed would seem to have adapted well to its harsh environment. But its adaptation has not allowed for the most recent onslaught on the desert. Almost all the Welsh's milkweed can be found within the Coral Pink Sand Dunes in Kane County in southern Utah. About half these dunes are within the Coral Pink Sand Dunes State Park, and the other half are on federal land managed by the Bureau of Land Management (BLM). The state park was created and funded, in part, as a playground for off-road vehicles. Though the Welsh's milkweed on BLM land is protected under the Endangered Species Act, adequate enforcement still requires local cooperation.

If sufficient protection of the wild habitat does not come soon, botanists may have to rely on artificial cultivation methods to keep the plant alive. But here, too, there is a problem. Though seeds taken from the wild plants by researchers from both Flagstaff Arboretum and Red Butte Garden and Arboretum in Salt Lake City are sprouting, the seedlings, unlike the parent plants, have narrow leaves and do not mature beyond this juvenile stage. The next step is to see if greenhouse plants grown from roots will produce seed.

West Indian Manatee

Trichechus manatus latirostris

Every plant and animal has something to teach human beings. Some of the lessons are buried deep in the creature's molecular structure, like the cancer treatment hidden in the bark and needles of the yew tree. Others are clear for all to see, like the gentleness of the West Indian manatee.

Males competing for the attention of a female in heat do sometimes engage in a ritual of gentle pushes and nudges. Otherwise these huge tropical vegetarians don't have a single aggressive behavior in their repertoire, though they have survived for about 50 million years. It was about that long ago that the manatee, whose closest living relative is the elephant, defied the evolutionary trend by crawling back into the sea.

But the manatee's nonviolent approach to the world has proved less effective in the last few centuries. The manatee is in the order Sirenia, which includes three other surviving members: the Amazonian manatee of South America, the West African manatee, and the dugong of Australia and the Indian Ocean. All are tropical and subtropical creatures, and all are herbivores, the only herbivorous marine mammals alive today. A fifth, and the largest species in the genus, was the Steller's sea cow. By the late 1700s, however, the Steller's sea cow, which grew to almost 30 feet in length, had been hunted to extinction by Russian sailors in the Bering Sea.

West Indian manatees spend their days lolling slowly about their river and ocean hangouts eating hydrilla, sea grasses, water hyacinth, and other plants (between 80 and 150 pounds a day) and trying to stay warm. Manatees are very susceptible to cold, despite their tough skin and thick layer of blubber. If they spend too long in water below 68°F, they risk contracting acute bronchopneumonia, a deadly disease for an animal unable to breathe through its mouth.

A far more serious threat to the continuation of this species is the proliferation of motorboats in the coastal Atlantic waterways. Manatees spend a lot of their time resting at the surface of the water and may be fatally injured when motorboats run over them. Most surviving manatees show scars where propellers have cut their backs. Attempts to control the speed and number of boats in primary manatee habitat are having only limited success. They are hard to impose and enforce, especially in Florida where water-sports are seen as essential to the local economy.

Whooping Crane
Grus americana

"The whooping crane, perhaps the most majestic of all our feathered hosts, has traveled the long trail into oblivion." So claimed a writer in the *Saturday Evening Post* in 1923. We now know he exaggerated, but not by much. Estimates show that earlier this century the population of whooping cranes had fallen to fewer than 25 individuals.

Today the main population of about 150 birds spends the summer in the Wood Buffalo National Park in Alberta and the Northwest Territories of Canada, and winters in the Aransas National Wildlife Refuge in southern Texas. The tallest American bird at four-and-a-half feet, the whooping crane presents a striking figure as it wades through the marshes and prairie potholes along its migration route, looking for frogs, snails, small rodents and fish, and carrion.

The cranes arrive at their wintering grounds in mid-November and remain there feeding on crabs and clams plucked from the salt marshes on the shores of the Gulf of Mexico. Once a crane reaches sexual maturity, usually four years of age, it finds a partner with whom it will mate for life.

The first signs of a new breeding cycle occur in mid-January with an elaborate bonding dance. First an individual, and then a pair, whether they are newlyweds or a long-term couple, pump their heads up and down, flap their wings, and jump in the air, their legs stiffened and toes thrust forward. When their movements are in unison and their two calls are as one call, the bonding appears complete.

In April the cranes, in twos and threes, head north to the breeding grounds. Here they build a nest beside the dense vegetation, the bulrushes and cattails that surround the shallow lakes. They will return to the same region, often the same nest, every year. The female usually lays two eggs, but both parents take turns during the month-long incubation and in keeping watch over the newborns for the first few weeks.

Within recent history whooping cranes have never been plentiful. Even before people began turning wetlands into farmland in the Midwest along the cranes' traditional migration route, it is believed there were never more than 2,000 birds. The whooping crane was one of the first species to be recognized as endangered, and one of the first to receive significant support. In 1975 a separate population of whooping cranes was established at Gray's Lake National Wildlife Refuge in Idaho, with sandhill cranes as surrogate parents. These birds created a migration route to the Bosque del Apache National Wildlife Refuge in southern New Mexico but have been disinclined to mate; the female whooping cranes preferred the male sandhills. In 1993 captive-bred chicks were introduced to single whooping cranes to see if they will bond with the adults and learn how to survive in the wild.

Nowadays, the greatest dangers the birds face are avian tuberculosis and power lines, but captive-breeding programs established at the Patuxent Wildlife Research Center in Maryland, the International Crane Foundation in Baraboo, Wisconsin, and the Calgary Zoo in Alberta have so far managed to mitigate successfully against the constant threats to the whooping crane's survival.

Winter-Run Chinook Salmon
Oncorhynchus tschawytscha

Turning sideways and flipping her tail, a female salmon digs a shallow depression, called a redd, in the gravel riverbed. She unloads perhaps 5,000 eggs while a nearby male releases a stream of sperm to fertilize her offering. A few more tail flicks buries the nest with gravel. Having spent three or four years roaming the northern Pacific Ocean, the pair have followed their natal stream's specific odor and mysterious navigational cues to return to where they were born, and where, having completed their instinctual mission, they will die.

Historically, this story played out thousands of times each year on California's largest river, the Sacramento, and its major tributaries: the McCloud, Pit, and Little Sacramento rivers. Every winter and spring, all these rivers once teemed with chinooks, or kings, the widest ranging of Pacific salmon.

The Sacramento is unique on the Pacific Coast in having year-round chinook spawning in four different runs, named for the season when adults migrate from ocean to river. The winter-run migration begins in December and peaks in March, with most spawning in May and June. Some chinook can weigh up to 100 pounds, but winter-run fish rarely exceed 20 pounds.

Since each run is a genetically distinct population, the winter run, the rarest of the four, became eligible for protection under the Endangered Species Act. California listed the winter run as endangered shortly before the federal government approved a threatened listing, which permits an "incidental take" of salmon by fishermen and by water-pumping stations used for irrigation.

The winter run's problems began after the building of Shasta and Keswick dams in the 1940s. The dams, lacking fish ladders, kept the salmon from reaching spawning grounds on other rivers and formed Lake Shasta, which covered Sacramento River spawning grounds. Cold water released below the dam allowed the significantly reduced run to persist, as the eggs and fry can only survive in water temperatures between 40°F and 56°F.

The opening of the 12-foot-high Red Bluff Diversion Dam in 1966 flooded six miles of salmon spawning grounds, creating a shallow summer sauna which cooked the eggs. More than 117,000 winter-run spawners made it past the Red Bluff Diversion Dam in 1969; 20 years later there were only 500. Since the winter-run chinook became threatened, the dam gates have been left down during spawning season, allowing the salmon access to the best remaining spawning grounds between the diversion dam and Shasta and Keswick dams to the north. This reach has been declared critical habitat for the chinook.

Recent years have seen a record-low run in 1991 of 191 fish, a run of 1,180 in 1992, and a projected run of 350 for 1993. The smaller the run becomes, the less likely it will preserve the genetic variability that has helped the salmon overcome previous environmental obstacles.

To bolster the population, the Coleman National Fish Hatchery in Anderson, California, has been propagating and releasing winter-run chinook. Half of the 1991 run was estimated to consist of hatchery-reared salmon. The Bodega Marine Laboratory in Bodega Bay, California, and Steinhart Aquarium in San Francisco have joined the effort to rear captive winter-run chinook and release offspring to the wild. Perhaps one day, the Sacramento will once again teem with salmon.

Wood Bison
Bison bison athabascae

Humans have hunted bison for at least 12,000 years, since both entered North America via a land bridge across the Bering Sea. But the continent's largest living land mammal has been pushed to the edge of extinction in just the past century. Railroads sealed the fate of the vast herds of plains bison that once roamed the frontier. Passengers shot them from passing trains, and one railroad conductor claimed that it was possible to walk for 100 miles beside the tracks without stepping off a bison carcass.

Never as numerous as plains bison, the wood bison of Canada's boreal forests also came into conflict with the ambitious expansion of humans. Athapascan natives hunted wood bison, but not to the extent that traders slaughtered bison in the late 1800s for meat and hides. By the turn of the century, only about 300 wood bison had survived the hunters and several unusually severe winters.

Taller, heavier, and with a more angular hump than its plains relative, the wood bison has been classified as a separate subspecies. Some scientists maintain that plains and wood bison, despite their physical and ecological differences, are genetically too similar to be split into two subspecies.

In 1922, the Canadian government established Wood Buffalo National Park in Alberta and the Northwest Territories to protect the last wood bison. Against the protests of biologists, in the 1920s more than 6,000 plains bison were transported from southern Alberta and released in the park. Outnumbering wood bison four to one, the transplants interbred with the wood bison herds and infected them with tuberculosis and brucellosis. Wood bison were thought to have vanished until a chance aerial survey in 1957 discovered a small herd in a remote corner of the park. Forty-one of these animals became the founders of new populations, but the fate of the park's remaining diseased bison is in question.

The goal of restoring herds began in 1963 with 16 animals at the Mackenzie Bison Sanctuary in the Northwest Territories. Now numbering 2,000, this is the last large, free-roaming herd. In 1965, 21 bison were relocated to Elk Island National Park in Alberta, where this photograph was taken. Two other recently established herds in the Yukon and Manitoba do not yet have the 200 to 300 bison needed to make a viable herd. Optimistic about the future, the Canadian government changed the wood bison's status in 1988 from endangered to threatened. If herds can also be released in Alberta's large tracts of habitat, then the wood bison may return to its former grandeur.

Wood Stork

Mycteria americana

The warm, shallow water where the wood stork wades is murky and tangled with vegetation. The stork even adds to the murkiness, stirring up the mud by pumping its foot up and down a few times prior to each step. The movement startles fish out from the cover of the plants. Sometimes, for the same effect, the stork flicks opens its wings, briefly casting a shadow over the water. Unlike other wading birds which depend on their keen sight to catch their prey, the wood stork relies on feel, on "tacto-location." It keeps its bill in the water all the time, slowly moving it to and fro until it touches something that moves. Immediately—actually in 25 milliseconds, one of the fastest vertebrate reflexes known— the wood stork closes its bill. Its catch is usually small fish about two inches long, but crayfish, frogs, and aquatic insects may also fall victim to its lightning reaction.

The wood stork, the only stork native to North America, is a large bird, about 40 inches tall, with a distinct head and neck, bald and gray, and an inward-curving bill. After an elaborate courting ceremony with much head bobbing and preening, followed by copulation, a pair builds a flimsy platform atop a tall cypress or mangrove tree. The wood stork is a gregarious creature and nests in colonies of up to a few hundred pairs. As many as 25 pairs might occupy a single tree. The eggs, from two to five, are laid one to two days apart. If there is a food shortage, clearly the oldest—the largest and strongest— chick gets the lion's share. This way nature improves the likelihood that at least one or two offspring should get by even in a bad year.

Lately, there have been too many bad years. The survival of the wood stork depends upon the reliability of its food source. A nesting pair may require 440 pounds of fish during the breeding season. As land is drained and wetlands are broken up, the deep holes (often dug by alligators) which historically ensured that fish populations would survive the winters will likely dry up. Although the wood stork population is relatively healthy for an endangered species of bird, and few people would dream of harming them, the inexorable demand for residential development and agricultural land may reduce the wood stork's habitat below the critical mass it needs to prevail.

Wyoming Toad

Bufo hemiophrys baxteri

The Wyoming toad is a glacial relic. As the earth cooled during the last ice age, the toad's ancestors moved south ahead of the advancing glaciers. And when the glaciers retreated, this population of amphibians remained isolated around a series of ponds in the Laramie Basin in southeast Wyoming. Over the next 12,000 years, the Wyoming toad evolved into a distinct subspecies, with a unique crown atop its head.

By 1986, the toad population was down to one small pond, Mortenson Lake in Laramie Basin. There, on a good year, there might be 50 adults, but it was clear to biologists that, if the year were a particularly bad one, with severe drought or an unusally warm, or cold, winter, one spring there might be no toads at all.

Fortunately, one of the toad's neighbors was the black-footed ferret (page 217), whose precarious status together with its charismatic charm had sprung loose funds under the Endangered Species Act to establish a captive-breeding program. With facilities and expertise so close, it seemed wise to hedge against the distinct possibility of the toad's demise.

There were many skeptics, because previous captive-breeding efforts had produced few, if any, toads that survived in the wild. But, armed with some innovative techniques, scientists felt it would be worth the gamble. First, they ensured strict sanitation in the toads' captive environments. Second, they created a diet of especially small things for this very small toad—tiny crickets, for instance, and larder beetle larvae. Third, they developed a climate for hibernation that allowed the toads to survive with their breeding abilities intact. When the time came for the toads to hibernate, they were transferred to a special refrigerator where they could burrow down into the sand and spend four to five months at temperatures down to 40°F and with the humidity around 80 percent.

But it doesn't work to put healthy captive-bred toads back in the wild and expect them to breed. Somewhere between the egg being laid and the tadpole emerging, the animal may have already imprinted on its breeding site.

So in 1992 adult toads from the Sybille Wildlife Research and Conservation Education Center were placed in a caged community on the edge of Lake George in the Hutton Lake National Wildlife Refuge in southeast Wyoming. The toads bred, the eggs hatched, and now biologists wait to see if sufficient offspring will survive disease, harsh weather, predators such as fish and gulls, and potential problems from inbreeding. Not until there are five or six sites with breeding populations of 200 or so animals will biologists feel that the toads are out of the woods and back in the wild.

Photographers' Note

DAVID LIITTSCHWAGER

AN AUDIENCE WITH AN ENDANGERED PLANT or animal is not easily obtained. Fortunately, our previous work photographing endangered species, our association with the California Academy of Sciences, and the skillful persistence of our production coordinator, Beth Sudekum, enabled us to gain access to the people, plants, and animals that made these photographs possible. The cooperation we received from the people who care for endangered species was extraordinary and crucial to the making of this book.

Ash Meadows sunray

Beth looking for Stephens' kangaroo rats

Among the situations we faced were the projected bloom times of the flowering plants, the seasonal habits of the animals, geographic distances, human schedules, and weather. Beth choreographed this juggling act and directed it via midnight telephone calls and any available fax machine. Over a period of 27 months, we zigzagged across the country chasing down plants whose bloom times defied botanists' best predictions, and traveled great distances to arrive at opportune times in animals' life cycles. We went to Gainesville, Florida, to catch the emergence of a Schaus swallow-tail butterfly, and to Duxbury, Massachusetts, to photograph piping plovers when they were sitting on their eggs.

The pace of the project was hectic. When we weren't photographing, we spent most of our time on interstate highways and in bland motel rooms. We outfitted a 20-foot-long stepvan, the kind used for bread deliveries, with the equipment needed to photograph anything from a tiny Tooth Cave spider to a 2,000-pound wood bison. The van was loaded with rolls of seamless paper; bolts of black velvet, black flannel, and canvas; lighting equipment and grip equipment; black-and-white, color transparency, and Polaroid film; lumber, foam core, fill cards, acrylic, and glass for constructing studio enclosures; tools; an electric generator; and several boxes of miscellaneous items to see us through almost any eventuality.

Our photographic technique was designed to be as noninvasive as possible while maintaining a stark background. When photographing small animals and plants, extra care was taken to avoid overheating them with the modeling lights. For the Attwater's prairie chicken, we had to use an enclosure familiar to the bird. We constructed a studio from materials regularly used for the cages of young birds, modified the shape to accommodate the camera, lighting, and background, and left enough room for the bird to feel comfortable.

Key tree cactus

Each species presented special challenges. The largest subject was the Key tree cactus. We wove a black velvet background behind the cactus and in front of the tangle of other plants in the dense hammock. The smallest subject, the Tooth Cave spider, was photographed on a piece of glass that could be moved to position the spider in the camera's frame. We needed an enormous expanse of black velvet to photograph the wood bison. Lacking any form of persuasion we waited patiently for four days for the bison to walk in front of our carefully arranged backdrop. Finally, one of the biologists explained that if you come near a wood bison it turns to show its profile

Although the circumstances under which we met these animals and plants are deeply regrettable, each creature asserted, by its continuing presence, by its unique way of being, a voice of hope.

TECHNICAL DETAILS

Victor Hasselblad Inc. supplied the cameras, a 553ELX and a 200FCW, with lenses ranging from a 25mm Luminar to a 500mm Tele-Tessar. The lens used most often was a 135mm S-Plannar on an automatic bellows.

We used either Balcar or Norman electronic flashes to light all of the animals except the wood

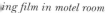
ing film in motel room

Welsh's milkweed

Susan and David in Wood stork studio

and impress upon you how big he is. So one of us approached the animal at 90 degrees to the camera and, on cue, the bison turned toward the lens.

Eureka Valley evening primroses live on the sand dunes at the southern end of Eureka Valley, California, where the wind almost never stops blowing. Their tissue-paper-thin petals flutter in even the slightest wind. We placed a small aquarium over the plant to still it and photographed it through the glass.

We used a 500mm lens for the California condor, the piping plover, and an adult Florida panther. The increased working distance was needed for these animals, but for the other pictures, as in traditional portraiture, we preferred to be close to our subjects.

When we photographed the American burying beetle, our main problem was the lighting. These beetles are quite something to behold once you get past the smell of the carrion on which they live and breed. They are shiny and hard and very black, with bright orange spots and huge black mandibles. Any black object on a black background is difficult to photograph. It's especially hard to light a spunky beetle whose face is only half an inch across. This kind of challenge was one of the many pleasures of this work.

bison, which was lit by the sun. For fieldwork inside or close to the van, we ran the Balcar flash units with a gasoline-powered generator. If we were at some distance from the van or in difficult terrain, we employed battery-powered Norman 200c electronic flashes. The plants were lit by the sun whenever possible. White or silver reflector cards were often used to fill shadows. If there was not enough light, or too much wind, we used electronic flash. We determined the exposure with a Minolta Flash Meter III, except when doing so would disturb the subject. In such cases we used Polaroid Type 669 or 664.

Most of the color photographs were taken on Kodak Ektachrome 64X. The black-and-white film was Kodak Plus-X. Twelve hundred rolls of color film were processed E-6 by The New Lab in San Francisco. One thousand rolls of black-and-white film were processed in Agfa Rodinal at the California Academy of Sciences. For the black-and-white printing, we used either a condenser head or a cold-light head on a Simmons Omega D2 enlarger with an 80mm Rodenstock APO-Rodagon-N lens. The black-and-white prints were made on Agfa Insignia paper using Agfa chemistry. Some of the prints were toned in selenium to increase the contrast and density of the shadows.

The Endangered Species Act

THE PURPOSES OF THIS ACT ARE TO PROVIDE MEANS WHEREBY THE
ECOSYSTEMS UPON WHICH ENDANGERED SPECIES AND THREATENED
SPECIES DEPEND MAY BE CONSERVED.

All the species in this book have one thing in common:
each occupies a place on the Federal List of Endangered and
Threatened Wildlife and Plants. This list is at the core of the
Endangered Species Act; it is the fulcrum around which the
provisions of the act are committed. The Endangered Species
Act (ESA) represents the nation's most determined effort to
take responsibility for preserving its precious biological diversity.
If the act continues to be honored, history will acknowledge its
passage as a turning point in the relationship between people
and the plants and animals with which we share our world, and
on which we depend for food, medicines, companionship, and
the climatic conditions that make life possible. By offering strict
federal protections to the species that are included on the list,
the government has drawn a philosophical line which it will not
allow human pressures to encroach. That line is extinction.

In both its scope and its irreversibility, *extinction* is the
most frightening, most conclusive word in our language. When
a species has been made extinct, not only have all its individuals
died, but the possibility of any such individuals ever existing
again has been annulled. It is against the finality of extinction
that the designations "endangered" and "threatened," which are
used throughout this book, draw their significance.

When lawmakers passed the ESA in 1973, most of them
did not recognize the dimensions of the extinction crisis that
was quietly mushrooming out of control. They had in mind
keeping a few celebrated species from crossing into oblivion:
the bald eagle, the grizzly bear, the whooping crane. But the act
was written in all-inclusive language, and once a species was
listed, that plant or animal's preservation became the primary
criterion when considering any alteration of its natural habitat
for whatever purpose. If an exception was made in one case, an
exception could be made in every case. The Endangered
Species Act was originally designed to allow for no exceptions.

Two major consequences stem from listing. One is the
preservation of a listed species' habitat. This is particularly true
of fauna as the act prevents private landowners from carrying
out any actions that would result in the death, or *take,* of a listed
animal. There are no similar controls over private land on
which a threatened or endangered plant might be growing.
However, if a listed animal or plant is found on public land, it is
given full protection.

The second consequence of listing is to concentrate
efforts toward the species' recovery. The likelihood of success
often depends on the resources the U.S. Fish and Wildlife
Service is willing or able to provide. The status of about 10
percent of endangered and threatened species has improved
since they were listed. Thirty percent are thought to be stable,
but the populations of nearly 40 percent have continued to
decline. The fate of the remaining 20 percent is unknown; at
least 2 percent have gone extinct.

The animals and plants in this book represent about an
eighth of the species currently listed as either endangered or
threatened. A further 3,000 animals and plants have been
identified as candidates for protection, but when it became
clear how broad were the ramifications of the listing of a
species—residential, commercial, and agricultural development

could be stymied, forests could not be cleared, roads and dams
could not be built—there was great reluctance by various vested
interests to include new species.

Once a candidate species has been proposed for listing,
the process of evaluation, overseen by the Fish and Wildlife
Service, is very methodical and is based on such criteria as the
magnitude of the threat and the species' taxonomic
distinctness. Nevertheless, to become a candidate for listing a
species must often have the focused attention of a group of
dedicated biologists willing to complete reams of forms full of
meticulously gathered scientific data. And, once listed, if a
species is to receive sufficient funding for its habitat to be
adequately preserved, it must have a certain amount of political
clout, and it helps a lot to have some popular appeal.

The seed of what would become the most serious test of
the act came just as the legislation was being passed. An
environmental impact report for a projected dam on the
Tennessee River discovered the existence of an endangered
fish—the three-inch snail darter—and it appeared the dam's
construction could jeopardize the fish's survival. As a result of
the controversy, amendments were enacted in 1978 that would
enable an interagency exemption committee to review and
possibly overrule the original provisions of the act. Its members
must feel that the consequences of enforcing the act outweigh
the benefits of a particular development. To many people's
surprise, the committee ruled in favor of the snail darter. To
bear the responsibility for extinction weighs heavy.

The Endangered Species Act was not drafted to carry the
country's conservation standard. It was designed to catch the
few, rare species that slipped through other protective
legislation. Unfortunately, the ESA is the only environmental
law with real teeth. Only when a species is facing extinction does
it come within the auspices of the act, and by then it is already
late, in some cases too late. Some species go extinct even before
they are listed.

In the 20 years since the passage of the ESA, more and
more attention has focused on the significance of a species'
habitat. To try to separate, at least legally, the fate of a species
from proposed development projects, amendments in 1982
were designed to enable a species to be listed without
immediately designating its critical habitat. This means that the
wheels of protection can be set in motion with less hindrance
from specific development interests whose eyes may be targeted
on a particular project.

Recent proposals, if enacted, including the planned
National Biological Survey, will separate further the listing of a
species from the determination of significant biological areas.
These proposals will focus more on ecosystem management
than on the immediate fate of individual species. This approach
may go a long way to anticipate the biological consequences of
development and prevent species from becoming endangered.
Only if we reduce the polarization between those who would
reshape the land and those who give precedence to the natural
world can humanity live alongside the wealth of the earth's
biodiversity.

FINDINGS

The Congress finds and declares that

1. various species of fish, wildlife, and plants in the United States have been rendered extinct as a consequence of economic growth and development untempered by adequate concern and conservation;

2. other species of fish, wildlife, and plants have been so depleted in numbers that they are in danger of or threatened with extinction;

3. these species of fish, wildlife, and plants are of esthetic, ecological, educational, historical, recreational, and scientific value to the Nation and its people;

4. the United States has pledged itself as a sovereign state in the international community to conserve to the extent practicable the various species of fish or wildlife and plants facing extinction. . . .

5. encouraging the States and other interested parties, through Federal financial assistance and a system of incentives, to develop and maintain conservation programs which meet national and international standards is a key to meeting the Nation's international commitments and to better safeguarding, for the benefit of all citizens, the Nation's heritage in fish, wildlife, and plants.

PURPOSES

The purposes of this Act are to provide means whereby the ecosystems upon which endangered species and threatened species depend may be conserved.

DEFINITIONS

Endangered: A species in serious danger of becoming extinct throughout all, or a significant portion of, its range.

Threatened: A species that, although not presently threatened with extinction, is likely to become an endangered species in the foreseeable future unless conservation steps are taken.

Fish and wildlife: Any member of the animal kingdom, including mammals, fish, birds, amphibians, reptiles, mollusks, crustaceans, anthropods or other invertebrates and includes any part, product, egg, or offspring, or the dead body or parts.

Critical habitat: (i) The specific areas within a geographical area occupied by a species, at the time it is listed. . . . on which are found those physical or biological features (I) essential to the conservation of the species and (II) which may require special management considerations or protection; and specific areas outside the geographical area occupied by the species at the time it is listed that are essential for the conservation of the species.

(ii) Specific areas outside the geographical area occupied by a species at the time it is listed determined essential for the conservation of the species.

Species: Any subspecies of fish or wildlife or plants, and any distinct population segment of any species of vertebrate fish or wildlife which interbreeds when mature.

Take: To harass, harm, pursue, hunt, shoot, wound, kill, trap, capture, or collect, or to attempt to engage in any such conduct.

BASIS FOR DETERMINATIONS

The Secretary [of Interior or Commerce] shall make determinations [for listing] solely on the basis of the best scientific and commercial data available after conducting a review of the status of the species and after taking into account those efforts, if any, being made by any State or foreign nation . . . to protect such species.

The Secretary [of Interior or Commerce] shall designate critical habitat, and make revisions . . . on the basis of the best scientific data available and after taking into consideration the economic impact, and any other relevant impact, of specifying any particular area as critical habitat. The Secretary may exclude any area from critical habitat if he determines that the benefits of such exclusion outweigh the benefits [of inclusion], unless he determines, based on the best scientific and commercial data available, that the failure to designate such area as critical habitat will result in the extinction of the species.

The Secretary shall (A) conduct, at least once every five years, a review of all species included in the Endangered Species Act . . . and (B) determine on the basis of such review whether any such species should—(i) be removed from such list; (ii) be changed in status from an endangered species to a threatened species; or (iii) be changed in status from a threatened species to an endangered species. . . .

LAND ACQUISITION

The Secretary, and the Secretary of Agriculture with respect to the National Forest System, shall establish and implement a program to conserve fish, wildlife, and plants, including those which are listed as endangered species or threatened species. . . . The appropriate Secretary . . . is authorized to acquire by purchase, donation, or otherwise, lands, waters, or interest therein. . . .

COOPERATION WITH THE STATES

The Secretary is authorized to provide financial assistance to any State, through its respective State agency . . . to assist in development of programs for the conservation of endangered and threatened species or to assist in monitoring the status of candidate species. . . .

INTERAGENCY COOPERATION

Establishment of a Committee

There is established a committee to be known as the Endangered Species Committee which shall be composed of: (A) The Secretary of Agriculture; (B) The Secretary of the Army; (C) The Chairman of the Council of Economic Advisors; (D) The Administrator of the Environmental Protection Agency; (E) The Secretary of the Interior; (F) The Administrator of the National Oceanic and Atmospheric Administration; (G) The President . . . shall appoint one individual from each affected State.

EXEMPTION

The Committee shall grant an exemption [to the Act] if it determines that—(i) there are no reasonable and prudent alternatives; (ii) the benefits of such action clearly outweigh the benefits of alternative courses of action consistent with conserving the species or its critical habitat, and such action is in the public interest.

PROHIBITED ACTS

With respect to any endangered species of fish or wildlife listed . . . it is unlawful for any person subject to the jurisdiction of the United States to (A) import any such species into, or export any such species from the United States; (B) take any such species within the United States or the territorial sea of the United States; (C) take any such species upon the high seas; . . . (G) violate any regulation pertaining to such species or to any threatened species of fish or wildlife listed. . . .

With respect to any endangered species of plants . . . it is unlawful for any person subject to the jurisdiction of the United States to . . . remove and reduce to possession any such species from areas under Federal jurisdiction; maliciously damage or destroy any such species on any such area; or remove, cut, dig up, or damage or destroy any such species on any other area . . . in the course of any violation of a state criminal trespass law.

Resource Guide

If you are willing to undertake the discipline and the difficulty of mending your own ways you are worth more to the conservation movement than a hundred who are insisting merely that the government and the industries mend their ways.

If you are concerned about the proliferation of trash, then by all means start an organization in your community to do something about it. But before—and while—you organize, pick up some cans and bottles yourself. That way, at least, you will assure yourself and others that you mean what you say. If you are concerned about air pollution, help push for government controls, but drive your car less, use less fuel in your home. If you are worried about the damming of wilderness rivers, join the Sierra Club, write to the government, but turn off the lights you're not using, don't install an air conditioner, don't be a sucker for electrical gadgets, don't waste water. In other words, if you are fearful of the destruction of the environment, then learn to quit being an environmental parasite. We all are, in one way or another, and the remedies are not always obvious, though they certainly will always be difficult. They require a new kind of life—harder, more laborious, poorer in luxuries and gadgets, but also, I am certain, richer in meaning and more abundant in real pleasure. To have a healthy environment we will all have to give up things we like; we may even have to give up things we have come to think of as necessities. But to be fearful of the disease and yet unwilling to pay for the cure is not just to be hypocritical; it is to be doomed. If you talk a good line without being changed by what you say, then you are not just hypocritical and doomed; you have become an agent of the disease.

WENDELL BERRY
A Continuous Harmony

There are many ways to learn more about endangered and threatened species and to become involved in the preservation of these species and their ecosystems. This resource guide includes some of the many conservation organizations and advocacy groups relevant to the subject of this book. Also listed are institutions that maintain and display North American species and libraries that focus on species preservation and biodiversity.

ORGANIZATIONS

Bat Conservation International
Austin, TX
Founded in 1982 by Merlin Tuttle to promote conservation needs of bats worldwide.

California Native Plant Society
Sacramento, CA
Seeks to preserve California's native plants; publishes magazine and newsletter, and an inventory of rare and endangered species.

Canadian Wildlife Federation
Ottawa, Ontario, Canada
Largest conservation organization in Canada; publishes species recovery plans and produces Project Wild and Habitat 2000, educational programs for youths.

Canadian Wildlife Service
Environment Canada
Ottawa, Ontario, Canada
Oversees two committees: Recovery of Nationally Endangered Wildlife (RENEW) provides recovery plans for endangered species; Committee on Status of Endangered Wildlife in Canada (COSEWIC) designates endangered species, maintains lists of species, and produces status reports. Environment Canada has branch libraries across the country.

Center for Marine Conservation
Offices in Hampton, VA; St. Petersburg, FL; San Francisco, CA; Seattle, WA; and Washington, D.C.
Promotes protection of critical marine habitats and endangered marine species; produces three publications.

Center for Plant Conservation
Missouri Botanical Garden
St. Louis, MO
Only national organization dedicated exclusively to preventing extinction of native plants; includes network of 25 botanical gardens and arboretums that collect and maintain plants listed in National Collection of Endangered Plants.

Conservation International
Washington, D.C.
Works in 23 countries to conserve tropical rain forests and threatened ecosystems; publishes quarterly newsletter and fact sheets.

Defenders of Wildlife
Washington, D.C.
Dedicated to protection of native animals and plants in their natural communities; publishes magazine and newsletter; coordinates program that publishes wildlife viewing guides for each state.

Delta Waterfowl
(formerly North American Wildlife Foundation)
Deerfield, IL
Conducts research and field programs to bring back North America's waterfowl populations; publishes newsletter; maintains Delta Waterfowl and Wetlands Research Station in Manitoba, Canada, for training biologists for waterfowl conservation worldwide.

Desert Fishes Council
Bishop, CA
International group of 500 research scientists and resource managers working to preserve biological integrity of North America's desert aquatic habitats and associated life-forms; publishes annual symposium proceedings and government reports.

Earth Island Institute
San Francisco, CA
Develops innovative projects for preservation and restoration of global environment; publishes journal, other publications, and videos.

Fish and Wildlife Reference Service
Bethesda, MD
Sells hard and microfiche copies of approved recovery plans for endangered species, as well as copies of documents on fish and wildlife and management research.

Friends of the Earth
Washington, D.C.
Conducts research and education programs on such environmental topics as tropical deforestation, ozone depletion, and global warming; publishes magazine.

Gopher Tortoise Council
Florida Museum of Natural History
Gainesville, FL
Concerned with protecting and preserving upland habitat necessary for survival of gopher tortoise.

Great Bear Foundation
Missoula, MT
Promotes conservation of bears and their habitat globally; awards grants for educational work and research on bear ecology; publishes magazine and serves as information clearinghouse.

Hawaii Plant Conservation Center
National Tropical Botanical Gardens
Lawai, HI
Collects, propagates, and studies Hawaiian plants, especially endangered species; offers education programs; collaborates with botanical gardens and land management agencies.

International Crane Foundation
Baraboo, WI
Maintains research and educational facility that breeds and restocks crane populations.

National Audubon Society
New York, NY
Concentrates on protecting habitat for migratory birds, strengthening Endangered Species Act, restoring Florida Everglades, and conserving natural ecosystems for all wildlife; produces magazine and endangered species tool kit; has more than 500 local chapters.

National Fish and Wildlife Foundation
Washington, D.C.
Established by Congress in 1984, the foundation has awarded over 700 grants totaling over $85 million for national and international conservation projects.

National Parks and Conservation Association
Washington, D.C.
Dedicated to conserving national parks; publishes magazine, newsletter, and reports.

National Wildflower Research Center
Austin, TX
Committed to preservation and reestablishment of native flora; provides more than 200 fact sheets to the public.

National Wildlife Federation
Washington, D.C.
Educates and assists individuals and organizations to conserve natural resources and protect environment.

Natural Resources Defense Council
New York, NY
Dedicated to protecting natural resources and public health through litigation, advocacy, and research.

The Nature Conservancy
Arlington, VA
Protects environments by purchasing land; owns 1,300 preserves throughout U.S. and has habitat-protection programs in Latin America, Caribbean, and Pacific.

Nature Conservancy of Hawaii
Honolulu, HI
Preserves biodiversity by protecting habitats for endangered species; has 13 preserves on five Hawaiian Islands and programs in several countries including Indonesia and New Zealand; publishes newsletter.

North American Native Fishes Association
Philadelphia, PA
Encourages conservation and aquarial appreciation of native fishes; publishes newsletter, magazine, and updated lists of protected fishes in U.S. and Canada.

Peregrine Fund, Inc.
Boise, ID
Focuses on conserving birds of prey, with captive-release programs in 43 countries; operates World Center for Birds of Prey in Boise, a facility to breed the California condor.

Save the Manatee Club
Maitland, FL
Works with government to protect manatees and their habitat; sponsors wild manatee adoption program and supports manatee research; publishes newsletter and provides educational material.

Sierra Club
San Francisco, CA
Committed to maintaining world's remaining natural ecosystems and to restoring degraded ecosystems; publishes magazine and booklet listing all endangered and threatened species in U.S. and Canada; has 400 local groups in U.S. and Canada.

The Wilderness Society
Washington, D.C.
Dedicated to preserving public lands in U.S. inhabited by endangered species; publishes newsletter, quarterly magazine, and fact sheets.

The Wolf Fund
Moose, WY
Focuses on recovery of wolves in Yellowstone National Park; educates public about wolves and addresses concerns of ranchers and hunters; publishes newsletter.

The Xerces Society
Portland, OR
Devoted to protection of invertebrates, focusing on endangered ecosystems and global biodiversity hot spots; publishes books, magazine, and educational tools. Name taken from Xerces blue butterfly, first butterfly in North America to become extinct because of human interference.

ARBORETUMS, BOTANICAL GARDENS, AQUARIUMS, AND ZOOS
Aquarium of the Americas, New Orleans, LA
Arizona-Sonora Desert Museum, Tucson, AZ
Assiniboine Park Zoo, Winnipeg, Man., Canada
Bok Tower Gardens, Lake Wales, FL
Calgary Zoo, Calgary, Alta., Canada
Cincinnati Zoo and Botanical Garden, Cincinnati, OH
Columbus Zoological Gardens, Powell, OH
Desert Botanical Garden, Phoenix, AZ
Garden in the Woods, Framingham, MA
Holden Arboretum, Kirtland, OH
Honolulu Zoo, Honolulu, HI
Los Angeles Zoo, Los Angeles, CA
Lowry Park Zoological Garden, Tampa, FL
Metro-Toronto Zoo, Toronto, Ont., Canada
Missouri Botanical Garden, St. Louis, MO
Monterey Bay Aquarium, Monterey, CA
National Marine Fisheries Service, Galveston, TX
New England Aquarium, Boston, MA
New York Zoological Park, Bronx, NY
Northwest Trek Wildlife Park, Eatonville, WA
San Antonio Botanical Garden, San Antonio, TX
San Antonio Zoological Gardens and Aquarium, San Antonio, TX
San Diego Zoo, San Diego, CA
Sedgwick County Zoo & Botanical Garden, Wichita, KS
Steinhart Aquarium, San Francisco, CA
Tennessee Aquarium, Chattanooga, TN
Vancouver Public Aquarium, Vancouver, BC, Canada
Waimea Arboretum and Botanical Garden, Oahu, HI
World Center for Birds of Prey/Peregrine Fund, Inc., Boise, ID

LIBRARIES
Biodiversity Resource Center
California Academy of Sciences
Golden Gate Park
San Francisco, CA 94118

Canadian Museum of Nature
P.O. Box 3443, Station D
Ottawa, Ont., Canada K1P 6P4

Environmental Action Coalition Library/ Resource Center
625 Broadway
New York, NY 10012

Environmental Library of Sarasota
7112 Curtiss Avenue
Sarasota, FL 34231

Fish and Wildlife Reference Service
5430 Grosvenor Lane, suite 110
Bethesda, MD 20814
(for recovery plans of most species)

Hawaii State Library
Hawaii & Pacific Section
634 Pensacola Street
Honolulu, HI 96813

New York Botanical Garden Library
P.O. Box 10458
Bronx, NY 10458

NOAA Library and Information Services Division
Central Library
6009 Executive Boulevard
Rockville, MD 20582

U.S. Fish and Wildlife Service
Patuxlent Wildlife Research Center Library
Endangered Species Group
11510 American Holly Drive
Laurel, MD 20708

Further Reading

GENERAL

Barker, Rocky. *Saving All the Parts: Reconciling Economics and the Endangered Species Act.* Washington, D.C.: Island Press, 1993.
A journalist's account of how protection of endangered species and local economies intertwine.

Berry, Wendell. *What Are People For?* Berkeley, Calif.: North Point Press, 1990.
Collection of essays examining human culture and agriculture.

————. *A Continuous Harmony.* New York: Harcourt Brace Jovanich, 1970.
Collection of essays examining human culture and agriculture.

DeBlieu, Jan. *Meant To Be Wild: The Struggle to Save Endangered Species Through Captive Breeding.* Golden, Colo.: Fulcrum Publishing, 1991.
Narrative account of the efforts of biologists to breed animals in captivity and restore them to their natural habitat.

Durrell, Gerald, with Lee Durrell. *The Amateur Naturalist.* London: Dorling, Kindersley, 1982.
A detailed guide for becoming a naturalist, emphasizing starting in your own backyard.

Ehrlich, Paul R., and Anne H. Ehrlich. *Population Explosion.* New York: Simon and Schuster, 1990.
Outlines the environmental threat of overpopulation and proposes programs for change.

Favre, David S. *International Trade in Endangered Species: A Guide to CITES.* Boston: Martinus Nijhoff Publishers, 1989.
Describes purpose of Convention on International Trade in Endangered Species of Wild Fauna and Flora, text of treaty, and how participating countries have interpreted text; lists species covered.

Kohm, Kathryn A., ed. *Balancing on the Brink of Extinction: The Endangered Species Act and Lessons for the Future.* Washington, D.C.: Island Press, 1991.
Twenty-one essays tracing history of 1973 Endangered Species Act; concludes with call for strategy to focus on protecting ecosystems and biodiversity.

Leopold, Aldo. *A Sand County Almanac.* New York: Ballantine Books, 1949.
Environmental classic describing ecology of a sand farm in Wisconsin and human impact on America's wildlands.

Lopez, Barry H. *Desert Notes: Reflections on the Eye of a Raven* and *River Notes: The Dance of the Heron.* New York: Avon Books, 1990.
Essays on the landscape of America.

————. *The Rediscovery of North America.* New York: Random House, 1992.
Compares present environmental exploitation to atrocities of the Spanish conquistadores.

Lowe, David W., John R. Matthews, and Charles J. Moseley, eds. *The Official World Wildlife Fund Guide to Endangered Species of North America.* 3 vols. Washington, D.C.: Beacham Publishing, 1990.
Comprehensive reference on North American species federally listed as threatened or endangered; includes description of habitat, range, distribution, and efforts toward conservation and recovery.

Matthiessen, Peter. *Wildlife in America.* New York: Viking Penguin, Elisabeth Sifton Books, 1987.
History of the destruction of North American wilderness from first colonists through 20th century, with information about numerous species.

Middleton, Susan, and David Liittschwager. *Here Today: Portraits of Our Vanishing Species.* San Francisco: Chronicle Books, 1991.
Collection of photographic portraits of 36 endangered species living in California, with brief profiles of each species' habitat, characteristics, and status.

Myers, Norman. *The Sinking Ark: A New Look at the Problem of Disappearing Species.* New York: Pergamon Press, 1979.
Overview of threats to species and tropical forests and consequences of their loss; gives roles of multinational corporations and international aid agencies and measures that can be taken to protect species.

Norton, Bryan G. *Why Preserve Natural Variety?* Princeton: Princeton University Press, 1987.
Convincing rationale for preserving wild species and ecosystems.

Snyder, Gary. *The Practice of the Wild.* San Francisco: North Point Press, 1990.
Essays on freedom, wildness, goodness, and grace, using lessons from nature.

Steinhart, Peter. *California's Wild Heritage: Threatened and Endangered Animals in the Golden State.* San Francisco: California Department of Fish and Game, California Academy of Sciences, and Sierra Club Books, 1990.
Gives overview of biodiversity in California, the state with the most endangered species, and sketches the status of 102 endangered species.

Stone, Charles P., and Danielle B. Stone., eds. *Conservation Biology in Hawaii.* Honolulu: University of Hawaii Press, 1989.
Collection of essays on unique ecosystems of Hawaii and efforts taken to preserve biological diversity of islands.

Stone, Christopher D. *Should Trees Have Standing? Toward Legal Rights for Natural Objects.* Portola Valley, Calif.: Tioga Publishing, 1989.
Establishes ethical basis for assigning legal rights to natural objects like animals and plants.

Wilson, Edward O. *Biophilia.* Cambridge: Harvard University Press, 1984.
Introduces *biophilia* as a term for the affinity humans feel for nature, and makes the case that interest in all living things may be a necessary element of human nature.

————. *The Diversity of Life.* Boston: Harvard University Press, Belknap Press, 1992.
Describes how species evolved and became diverse, the five major cataclysmic events that disrupted evolution and diminished global diversity, and humanity's initiation of the sixth great extinction spasm.

PLANTS

Kimura, Bert Y., and Kenneth M. Nagata. *Hawaii's Vanishing Flora.* Honolulu: Oriental Publishing Company, 1992.
Guide to islands' rare endemic plants with short descriptions and color photographs.

Long, Robert W., and Olga Lakela. *A Flora of Tropical Florida.* Coral Gables, Fla.: University of Miami Press, 1971.
Nearly 1,000 pages with information for identifying plant species known to grow without cultivation in south Florida, with an overview of geology and plant communities of the region.

Luer, Carlyle A. *The Native Orchids of the United States and Canada.* New York: New York Botanical Garden, 1975.
Describes every species of orchid, with color photographs and line drawings.

Zomlefer, Wendy B. *Flowering Plants of Florida: A Guide to Common Families.* Gainesville, Fla.: Biological Illustrations, 1989.
Provides general summaries of plant characteristics and distribution, and economic products derived from the plants, with illustrations.

MAMMALS

Burton, John A., and Bruce Pearson. *The Collins Guide to the Rare Mammals of the World.* Lexington, Mass.: Stephen Greene Press, 1987.
Guide to 1,179 species of endangered mammals throughout the world, including descriptions of each animal, its biology and ecology, population, cause of decline, and conservation efforts.

Nowak, Ronald M. *Walker's Mammals of the World.* 5th ed. 2 vols. Baltimore: Johns Hopkins University Press, 1991.
Comprehensive reference on more than 4,000 different species, with photographs; includes description of habitat, geographic distribution, and characteristics of each species.

BIRDS

Doughty, Robin W. *Return of the Whooping Crane.* Austin: University of Texas Press, 1990.
Chronicles efforts to save whooping crane from extinction, providing a historical overview of research on the crane.

Ehrlich, Paul R., David S. Dobkin, and Darryl Wheye. *Birds in Jeopardy.* Stanford: Stanford University Press, 1992.
Outlines histories of extinct bird species and status of imperiled species; gives contemporary and historical ranges, ecological requirements, and current threats; illustrated with color paintings.

REPTILES AND AMPHIBIANS

Conant, Roger, and J. T. Collins. *Reptiles and Amphibians Eastern/Central North America.* 3rd ed. Boston: Houghton Mifflin Co., 1991.
A book in the Peterson Field Guide series intended for identification of specimens; uses color photographs combined with color illustration.

Grenard, Steve. *Handbook of Alligators and Crocodiles.* Malabar, Fla.: Krieger Publishers, 1991.
Offers general information about locomotion, biology, and reproduction of alligators and crocodiles, followed by detailed accounts of each subfamily, including its range in U.S.

Klauber, Laurence M. *Rattlesnakes: Their Habits, Life Histories, and Influence on Mankind.* Berkeley: University of California Press, 1982.
Abridged version of 1956 two-volume edition thoroughly describes status of rattlesnakes, their biology, populations, and bite; includes myths and folklore about rattlesnakes.

Pritchard, Peter C. H. *Encyclopedia of Turtles.* Neptune, N.J.: T.G.H. Publications, 1979.
Covers all living turtles species, turtle evolution, and fossil history.

FISH

Eschmeyer, William N., et al. *A Field Guide to Pacific Coast Fishes of North America.* Boston: Houghton Mifflin Co., 1983.
Describes species and behavior of fishes found from Alaska through Mexico.

Moyle, Peter B. *Fish: An Enthusiast's Guide.* Berkeley: University of California Press, 1993.
Describes evolution and classification of fish, their behavior, diversity, and different habitats, and also gives accounts of the decline of many fishes, such as the pupfish.

Page, Lawrence M., and Brooks M. Burr. *A Field Guide to Freshwater Fishes.* Boston: Houghton Mifflin Co., 1991.
Surveys the freshwater fishes found in North America.

INVERTEBRATES

Berebaum, May. *Ninety-nine Gnats, Nits and Nibblers.* Champaign, Ill.: University of Illinois Press, 1991.
Narrative accounts of the unique lives of various insects.

Evans, Glyn. *The Life of Beetles.* London: Allen and Unwin, 1975.
General overview of beetles.

Linsemaier, Walter. *Insects of the World.* New York: McGraw-Hill, 1972.
Exhaustive reference on insects; complete resource on their biology and ecology, extending to subclasses.

Acknowledgments

The California Academy of Sciences would like to thank all donors to the North American Endangered Species Project, including the following major supporters whose generosity helped create the project:

American Airlines	The Chronicle Publishing Company
Anonymous	Victor Hasselblad, Inc.
BankAmerica Foundation	Anne T. & Robert M. Bass
Mr. & Mrs. Peter B. Bedford	National Fish & Wildlife Foundation

Acknowledgments from SUSAN MIDDLETON and DAVID LIITTSCHWAGER

We want to express our gratitude to the many people and organizations who helped to make this project possible.

To the CALIFORNIA ACADEMY OF SCIENCES for believing in and supporting this project,
and for its long-standing commitment to the study of biodiversity
To ROY EISENHARDT, Executive Director, for his faith in this project and for creating a context
at the Academy in which it could be realized

To THOMAS EISNER for his enthusiastic support, for turning our eyes toward the littlest lives,
and for caring so passionately about the world

To EDWARD O. WILSON for saying yes

To BARRY LOPEZ for his continued interest and for accompanying us in the field

To JACK JENSEN and CAROLINE HERTER of Chronicle Books for wanting to publish this larger vision

To BETH SUDEKUM, Production Coordinator, for her powers of persuasion

To JOHN McCOSKER and the staff of Steinhart Aquarium for providing a home base for the project

To PAUL HOFFMAN for documenting the fieldwork on video

To LYDIA MODI VITALE for her courage and vision

To DIANE ACKERMAN for nourishment

To JODY GRUNDY for her ideas and open heart

To WILLIAM EMERY for making the first donation to the project when it was still an idea

To LEWIS COLEMAN and JOHN LARSON for their personal attention toward launching the project

To JEFFREY MEYER for his vote of confidence and help landing a big one

To JUDITH DUNHAM, editor, and to LUCILLE TENAZAS, designer, for their expertise and care

To MELISSA HARRIS and BROOK DILLON for generously donating countless hours helping with the film and prints

To BRENDA SANDBURG, CHARLOTTE SPANG, DANIELLE WILLIAMS, and AMY WILSON
for their enormous commitment toward researching species for the text

To ERNST WILDI and PETER POWER from Hasselblad for making equipment available to us, always,
and often on short notice

To ARSENIO LOPEZ, SAM HOFFMAN, DAN OSHIMA,
and all the staff of The New Lab for their support and taking special care processing our color film

To VOLKER and BERNT VON GLASENAPP at Just Film for finding the film we needed, always,
and getting it to us when we needed it, wherever we were

To KEONI WAGNER, Director of Corporate Communications for Hawaiian Airlines,
for waiving all baggage charges on 32 equipment cases

To CHRISTOPHER BROWN at Calumet/Balcar for providing lighting equipment

To LOUIS SHU at Ken Hansen Photographic for dispatching additional equipment to us in the field

To BOB SALOMON and ROGER BARTZKE from H. P. Marketing for providing the lenses we used in the darkroom

We are deeply indebted to those people who gave so generously of their time and expertise with each and every species we photographed.

Rhoda Aceano
Deb Adamson
Lauri Ahlander
Steven Ahlstedt
John Aiken
William Alexander
Frank Almeda
Bonnie Amos
Christine Anderson
Jon Andrews
Susan Arbuthnot
Karen Ardoin
Paul Arnaud
Johnny Arnett
Mark Bagley
Stephen Bailey
Steven Bailey
Richard Ball
Luis Baptista
Rodney Bartgis
Ken Beebe
Lloyd Beebe
Douglas Bell
Tim and Tanya
  Binder
Sandra Blum
Marlin Bowles
Marlee Breese
Lauren Brown
Marie Bruegmann
Bill Brumback
Dick Buchonis
Noel Burkhead
Charles Caillouet
Pat Callahan
Mary Capperino
Don Carlsmith
Joe Carrier
Karen Cebra
Frank Chapman
Dusty Chivers
Joseph M.
  Choromanski
Glen Christensen
Miyoko Chu

Mike Clark
Rick Clawson
Sheila Conant
Andrea Conley
Carolyn Corn
Buff Corsi
Monte Costa
Stephanie Costelow
Paul Cox
Jim Cunningham
Tom Daniel
Jaret Daniels
Mary DeDecker
Marcella De Mauro
Phil Dendel
Mark Deyrup
Dena Dickerson
Rob Doster
Robert Drewes
Liz Ecker
Bob Ehrig
Zella Ellshoff
Dawn Elmblad
Thomas C. Emmel
Andrew Engilis
William Eschmeyer
Don Falk
Tom Fenske
John Fitzpatrick
Gene Forbes
Eric Forsman
Elizabeth Forys
Craig Freeman
William Garnett
Sue Gawler
Edward and Laura
  Gerstein
William Gilmartin
Michael Glenn
Tom Goff
Doug Goldman
Mike Goode
Andrew Gordon
Terry Gosliner

John Graham
Terry Graham
Kerry Grande
Julie Greenberg
Harry Greene
Steve Gregory
Curtice Griffin
Michael Hadfield
James Hammer
Jane Hansjergen
John Harris
Susan Healy
Anne Hecht
Williard Heck
Scott Hecker
John Hewitt
Ian Hiler
Jim Hill
Bob Hobdy
Don Hoffman
John Holsinger
Brooks Holt
Frank Howarth
Ken Howell
Jerome Jackson
Judy Jacobs
Carl Jantsch
Doug Janz
Jack Jeffrey
John Jenkensen
Royce Jurries
Steve Kaiser
David Kavanaugh
Irving Kawashima
Hank Kiliaan
Kim Kime
Liz Kools
Andrea Kozol
Debbie Krista
Larry Kruckenberg
Margaret
  Kuchenreuther
Steve and Linda
  Kuntz
Don Kwiatkowski

Charles Lamoureux
Gus Lane
James Lane
Cherie Langlois
Shawn Larson
Jim Lewes
Kahikina Lidstone
Phil Lightfoot
Crista Lobue
Fred Lohrer
John Lucas
Jack Luschinsky
Peter Luscomb
Laurie MacIvor
Doris Mager
Bill and Dorothy
  Mandell
Stuart Marcus
Delores Marvin
Ed Maruska
Lucile McCook
Robert McElyea
Loyal Mehrhoff
Brien A. Meilleur
Bob Meinke
Scott Melvin
Tom Mendez
Janis Merritt
Therese Meyer
Denny Miller
Cynthia Mills
Steve Montgomery
Debbie Moore
Peter Moyle
Nora A. Murdock
Henry R. Mushinsky
Rob Nawojchik
Rob Nicholson
John-O Niles
Jim O'Brien
Ben Okaimoto
Wes Olson
David Orr
Charles Painter
Bob Parenti

Brian C. Parsons
Tom Paulek
Bruce Pavlik
Steve Perlman
Carl Peterson
Marshall Plumer
Jackie Poole
Lisa Potter-Thomas
Mary Price
Ann Prince
Linda Pritchett-
  Kozak
Barbara F. Pryor
Diana Pugh
Hugh Quinn
Diane Ragone
John Rampley
Peter Raven
Richard Reams
James Reddell
Rich Reiner
John Reuth
Hal W. Reynolds
Ron Reynolds
Bruce Rosenthal
David Rosenthal
Edward S. Ross
Bob Rowland
Stephen B. Ruth
Lex Salisbury
Lee Schoen
Mark Schwartz
Larry Serpa
Larry Shoemaker
Roger Shudes
Christina J. Slager
Norman Sletteland
Clio Smeeton
Bob Smith
Mike Smith
Roland Smith
Tim Smith
Barry Spicer
Paul Springer
Marty Stein

David and Beth
  St. George
Craig W. Stihler
Dale Sweetnam
Bill Tab
Michael Tewes
Terri Thomas
Tom Thorne
Robbin Thorp
Doug Threloff
Stephen Tim
Jean Tinsman
Bill Tolin
Tom Tucker
Wayne Tyndall
Darrell Ubick
Kevin van der
  Molen
Joe Vaughn
Kathryn Venezia
Loren Vigne
Jens Vindum
Susi von Oettingen
Keoni Wagner
Charles Wakida
Jack Wallace
Mike Wallace
Michael Wells
Dave White
Bob Whiteford
Greg Wieland
Dan Williams
Paula A. Williamson
Glen Woolfenden
Keith Woolliams
Wendy Worth
Kenneth Yates
Lloyd Yoshina
Una May Young
Carl Zimmerman

We warmly thank the entire staff of the California Academy of Sciences, who contributed to the project in a myriad of ways.
We would like to extend a special thank you to the following people.

Daphne Derven
Robert Drewes
Blake Edgar
Robert Glavin
Suzanne Gray
Keith Howell

Michael Kucz
Anne Malley
Anne Marie McKenna
John Norton
Mark O'Brien
Suzanne Paras

Dan Pledger
Judy Prokupek
Darryl Skrabak
Gordy Slack
Lori Smith
Donna Worthington

Many organizations graciously opened their doors to us. Their cooperation and assistance are deeply appreciated.

Amy Greenwell Ethnobotanical Garden
Aquarium of the Americas
Archbold Biological Station
Ash Meadows National Wildlife Refuge
Audubon Institute
Beacham Publishing
Biltmore Estate
Bishop Museum
Boca Chica Naval Air Station
Boston University
Brooklyn Botanical Garden
Busch Gardens Zoo
California Academy of Sciences
California Department of Fish and Game
California State University, Fullerton
Canadian Wildlife Service
Center for Plant Conservation
Chihuahuan Desert Research Institute
Cincinnati Zoo and Botanical Garden
Cochrane Wildlife Reserve
Coleman National Fish Hatchery
Columbus Zoological Gardens
Coral Pink Sand Dunes State Park
Desert Botanical Garden
Elk Island National Park
Flag's Pond Nature Park
Florida Game and Freshwater Fish Commission
Fossil Rim Wildlife Center
Garden in the Woods/New England Wildflower Society
Golden Gate National Recreation Area
Graham Breeding Facility, Point Defiance Zoo
Haleakala National Park
Hawaii Heritage Program, Nature Conservancy of Hawaii
Hawaii Department of Land and Natural Resources
Hawaii Division of Forestry and Wildlife
Holden Arboretum
Honolulu Zoo
Houston Zoo
Humboldt State University
Isis Oasis
Kansas Ecological Reserves, University of Kansas
Los Angeles Zoo
Lowry Park Zoological Garden
Maine Natural Heritage Program
Maryland Natural Heritage Program
Massachusetts Audubon Society

Miami Seaquarium
Mississippi State University, Starkville
Missouri Botanical Garden
Mystic Marinelife Aquarium
National Fisheries Research Center
National Key Deer Refuge
National Marine Fisheries Service
National Tropical Botanical Garden
The Nature Conservancy, All Regions
New England Aquarium
New York Botanical Garden
Northwest Trek Wildlife Park
Olympic Game Farm
Panaewa Rainforest Zoo
Pohakuloa Training Area
Sacramento Zoo
San Antonio Botanical Garden
San Antonio Zoological Gardens and Aquarium
San Francisco Zoological Gardens
San Jacinto Wildlife Area
Saratoga National Fish Hatchery
Sea Life Park
Sea World of California
Southwestern Texas State University
Sybille Wildlife Research and Conservation Education Center
Tennessee Valley Authority
University of California, Berkeley
University of California, Riverside
University of Florida, Gainesville
University of Hawaii, Kewalo Marine Laboratory
University of Kansas, R. L. McGregor Herbarium
University of South Florida
University of Texas, Texas Memorial Museum
University of Utah, Red Butte Gardens and Arboretum
U.S. Fish and Wildlife Service, All Regions
U.S. Forest Service
Waikiki Aquarium
Waimea Arboretum and Botanical Garden
Wallowa-Whitman National Forest
West Virginia Division of Natural Resources, Wildlife Resources
White Oak Conservation Center
Wilson's Creek National Battlefield
Wolf Haven International
World Center for Birds of Prey/Peregrine Fund, Inc.
Wyoming Game and Fish Department

Our deep thanks extend to the people whose friendship, talent, and enthusiasm have helped spirit the project along.

We are especially grateful to the following people for providing havens and extending hospitality to us during the photographic production.

Tom and Maria Eisner
Bill Alexander
Virginia and Richard Bodner
    at Inspiration
Patricia Burke
Betsy, Sarah, and Jesse Eisenhardt
Kirk Frederick and Michael Laughlin
Roy and Maude Gardner
Terry and Janet Graham
Gordon Granger
Terry and Jodine Grundy
Anne and Scott Hecker
Don Kwaitkowski
Jean and Terry Liittschwager
Sandra Lopez and Desert
Lucile McCook and Cliff Ochs
Jean Middleton
Harry Miller
David Orr
Dorothy Richardson at Ravenridge
Mary Richardson and Theodore
Chris Ritecz
Lex Salisbury and Lisa West
Demetrios Scourtis
Kaiti Scourtis
Christina J. Slager and Joseph M.
    Choromanski
Clio Smeeton
Roland Smith
Walter Sorrell
Craig and Chris Stihler
Tom Thorne
Ellen and Lew Turlington
Fran and Jack Williams
Laura and Bobby Wilson
Diane Wolff
Wendy Worth
Una May Young

Carol and Susan Anderson
Tom and Tamia Anderson
Richard Avedon
Bob Barclay
Wendell and Tanya Berry
Adam Block
Gita Bodner
Gus Bodner
Heather Bradley
Yolanda Cuomo
Suzanne Fritch
Heather Hafleigh
John and Jodi Hockey
Ruedi Hofmann
Deidre Kernan
Tom Marioni and the MOCA Group
Rodney Marzullo
Svante Mossberg
Chuck Naacke
Wes Nisker
Christina Rahr
Robert Seidler
Alison Silverstein
Mary Troychak

and to Suzie Rashkis for her ideas
and companionship

Index